Favela Tours

Tourism and Mobility Systems Set

coordinated by
Philippe Violier

Volume 6

Favela Tours

Building Otherness in the Favelas of Rio de Janeiro

Thomas Apchain

ISTE

WILEY

First published 2023 in Great Britain and the United States by ISTE Ltd and John Wiley & Sons, Inc.

ISTE Ltd
27-37 St George's Road
London SW19 4EU
UK

www.iste.co.uk

John Wiley & Sons, Inc.
111 River Street
Hoboken, NJ 07030
USA

www.wiley.com

Library of Congress Control Number: 2023943508

British Library Cataloguing-in-Publication Data
A CIP record for this book is available from the British Library
ISBN 978-1-78630-767-5

Contents

Introduction

Since the 1990s, tourists, mostly European or North American, have visited the favelas of Rio de Janeiro. In small groups, they follow guides through what were once places that tourists carefully avoided. They learn about their history and try to understand how their inhabitants live there. Favela tours really exploded in the 2010s, before taking a lesser place at the end of the decade. It was during their peak that this research, conducted between 2012 and 2018, was carried out, based on repeated ethnographic observations of the tours and interviews conducted with their protagonists: tourists, guides and inhabitants. The aim of this book is to present this investigation in order to lead a reflection on the development of tourism in the favelas, representative of the contemporary modes of construction and consumption of otherness.

Indeed, while the favelas have become a place for tourist mobility, it is first of all because it constitutes an otherness now valued by those who travel in Rio de Janeiro. Although tourism is now common, it continues to be thought of as a novelty by those who practice it, or at least as an excursion “off the beaten track”. In this respect, tourism in the favelas is not an isolated phenomenon. On the one hand, it is linked to the emergence of tourism in other places that have the particularity of having been initially thought of as non-touristy, either because of a lack of interest or more often because of the difficulty of getting there and/or the danger. Here, the case of favelas is close to other tourism practices that some researchers (Frenzel and Koens 2012) have gathered under the label of slum tourism, to refer to the emergence of tourism in poor neighborhoods of major metropolises around the globe. Although “slumming” has older origins, the phenomenon seems, in these proportions, characteristic of the 2000s and 2010s. On the other hand, the growth of tourism in the favelas is part of a tendency that can be found in other practices, including alternative practices in typical tourist sites, and to which correspond strategies used by both tourists and their hosts. For tourists, the promise of a visit “off the beaten track” is linked to strategies of distinction and the acquisition of a symbolic prestige. For the hosts, the valorization of the non-touristic element allows them to distinguish objects, places or practices in order to insert themselves – not without paradox – into the tourist economy.

In other words, the ethnography of favela tours allows us to observe the emergence of a contemporary form of tourism, while including it in the cultural mechanisms that have long structured the practices of leisure mobility oriented towards the discovery of the Other. Indeed, it is a specific form of tourism, a tourism of the Other, that will be discussed here. While the analyses developed in this book will sometimes go further than the case of the favela tours, they will be limited to this form that, by the intercultural situations that it brings about, corresponds best to the epistemological specificities of anthropology. Tourism is a shape-shifting phenomenon that affects all areas of social life, which is why it is considered by some to be a "total social fact" (Cousin and Réau 2009), and in any case it raises several questions for all social sciences. For anthropology, it is a matter of studying its cultural dimensions and analyzing it as a practice that is above all part of an "economy of otherness" (Cousin and Apchain 2016; Cousin 2018). Anthropology is therefore faced with the need to make a choice. Either it places itself on the tourists' side, observes their ways of doing and thinking, questions their representations, listens to their stories. Or, it places itself on the other side, that of the hosts, and observes how tourism fits into the local culture. In short, it either deals with excursions or incursions. Finally, we should mention that anthropology has been able to place itself at the level of intermediaries, in particular guides (Doquet 2009), protagonists between two worlds. In any case, it seems complicated for anthropology to study all the protagonists at once, mainly because it is based on a method – that of ethnography and participant observation – which implies that it tries to see through the eyes of a specific cultural group.

In its young history (Leite and Graburn 2010), the anthropology of tourism seems to have mostly taken on the task of studying tourism from the perspective of host populations. In doing so, it ensures anthropology's continuity, not only because of tradition and anthropologists' appetite for the distant, but also because it has made the study of exogenous groups to the researcher an essential point of its method, that of decentering. More often, however, tourism has come to find the anthropologist in its "field", and the latter has tried therefore to interpret the ways in which the locals, whether or not they became hosts, interpreted the phenomenon, interacted with the tourists, understood their motivations or simply learned to live with tourism. From this host-centered perspective, the study of cultural change emerged as a dominant issue, at first, in a negative and deterministic way, and then, with a growing awareness of all the negotiations, games and syncretisms caused by tourism.

However, there are still many lessons to be learned from an anthropology of tourists, and this is the approach that this book advocates. Making the ethnography of tourists is a difficult task for many reasons (partly because of their permanent mobility) and may seem to depart from the principles of discipline, which advocates a decentering that a geographical and cultural distance is supposed to ensure. However, the tourist poses this problem regardless of the chosen approach and imposes a need for reflexivity that was often lacking at the time of its arrival at the ethnologists' doors. More on the side of an anthropology of the West than of an anthropology of the distant, although the field itself is geographically distant, this research is therefore concerned with tourists, their representations, the symbolic motivations by which they behave and interpret their practice, and the place that the favela occupies in their relationship to otherness. It is therefore intended to reflect on the cultural dimensions of the practice of tourism and not on those of its reception or impact, although, as far as these questions are concerned, understanding tourists also makes it possible to understand how tourism makes local upsurges.

The method chosen follows this orientation and attempts to adapt to the particular type of respondent that the tourist embodies. The analyses that this book seeks to develop are based primarily on the content of the excursions, the discourse that is expressed and received within the tours, the routine of their development, the events that may occasionally occur, the explanations of the guides to the tourists and the accounts that those who participated give about the visit. It is the desire to collect these accounts that constitutes the first invitation to go beyond the traditional framework of the "field" in anthropology, often thought of as a unity of place. Indeed, in order to understand the experience of tourists, it is sometimes necessary to situate its analysis in a wider framework than that of the practice itself, in this case the favela tour. To understand the value and uses of the tourist experience, it is necessary to be able to talk to tourists and/or see how they recount it after their trip, back in their own society. Here, the interviews were not systematically conducted during and after the trip, but I tried to remain attentive to this type of data. On several occasions, I will therefore refer to interviews that took place outside of Brazil (mainly in France and a few in the United States), with people who had visited a favela in the past. These interviews are essential for a tourist-centered approach, as a fundamental part of what drives travelers to visit a favela is played out after the visit, even more so when the story is told after returning home in a way that often greatly modifies the experience.

Some of the observations described in this book also come from moments outside of the tours, as a result of the relationships formed with the tourists encountered during the favela tours and sometimes extended after their trip. In some cases, I will include descriptions from stays in youth hostels, which allowed me to experience the daily life of travelers for several weeks. Finally, while this investigation is centered on the tourist, it nevertheless grants an important place to the main interlocutor, the guide. Indeed, the guide is a key character, insofar as he controls the essence of the representation of the favela for tourist purposes. As I will show, the guides' success depends largely on their ability to surprise, while presenting the favela in terms that correspond to Western criteria of valuing otherness. As the guides are situated between the world of the tourists and the favela, the analysis of their path, activity and discourse is indispensable. Other characters who gravitate around the excursions will be mentioned, but will not be the subject of an in-depth analysis.

This investigation took place from 2012 to 2018, over a period that corresponds to the rise of the favela tours. The latter have existed for 30 years in similar forms, although the phenomenon as a whole has had a more marginal existence in quantitative terms. The rise of the favela tour, between 2008 and 2016, was not only due to the slow maturation of tourist practices that appeared in the early 1990s. It was above all the consequence of the emergence of Western representations valuing the favela and of a context marked by local initiatives that took advantage of (but more generally dealt with) the evolution of its image and with the appearance of tourism on a greater scale.

1

The Invention of the Tourist Favela

"For me, I've always wanted to see this. It is the first thing that I've done." Léo[1], a 26-year-old Frenchman, had hardly arrived in Rio before he found himself exploring a favela. He was alone in a group of a dozen tourists, all French, enjoying a caipirinha in Vila Canoas, which was the highlight of the visit organized by Favela Tour, the eponymous pioneer of the tours organized in the favelas. Léo and I were talking. He had on a T-shirt with the colors of Brazil marking "Rio de Janeiro":

> *Léo*: They lost my suitcase at the airport.
>
> *Me*: Oh boy, and have you just got here?
>
> *Léo*: Yes, yesterday.
>
> *Me*: And did you come here on the first day?
>
> *Léo*: Well yes, I really wanted to do this, and then as, the weather was bad today, I came here (Conversation with Léo, French tourist visiting Rocinha, July 2012).

Léo's example is not a common one. While, in 2012, travelers came to some favelas in large numbers every day, it was rarely the first day of their stay. Most of them came after visiting the city's attractions – Corcovado and its Christ the Redeemer, Sugarloaf, Maracanã – and sometime after enjoying the beaches of Copacabana or Ipanema. In short, if they did a favela tour, it was most often towards the end of their trip or, indeed, when it rained. Tourists who, like Leo, make visiting

1 This work uses pseudonyms, except in cases where I have been given permission to use real names.

a favela a priority exist, but are rare. Later, as we left the favela, I asked Léo again about his visit. He said:

> For me, this is Brazil. I like Brazilian music a lot and, I don't know why, personally, it refers more to that [the favela], and the movies too. In fact, I have the impression that all I know about Brazil is the favela. Even if I didn't see everything, it was still good (Léo, French tourist visiting Rocinha, June 2012).

What Léo points out is the existence of a set of images, representations and ideas of what the favela is – a set of things that are not systematically shared in such an elaborate and positive way, but which constitute an essential piece in explaining the tourists' attraction to the favelas. Without being the exclusive factor of attraction, what is sometimes called the imaginary[2] must be questioned. Its place as the driving force behind the attraction to destinations is obvious and, for some, it is its impact that characterizes tourism as a matter of recognition (Urry 1990), aiming at "noticing how much the originals resemble their copies" (Augé 1997, p. 23). Indeed, it is certain that this imaginary has a fundamental place as a medium for practices, even though it is not shared by all in the same terms and even though, in the end, tourists like Léo only rarely "see" the favelas as they already "know" them.

1.1. The favela in the imaginary

The term "favela", as we shall see, refers to a polysemous category, defined above all by a set of oppositions, the most general of which is the division between *favela* (or *morro*) and *asfalto*. These oppositions articulate at the local scale a fragmentation of the urban space and a stigmatization of the favelas. While belonging to the category of favela can certainly be activated in a positive way in the assertion of their cultural specificities by its inhabitants, it is most often the source of a tenacious devaluation that pays little heed to the extraordinary diversity of spaces qualified by the term "favela". However, it appears that the representations associated with the favela vary according to whether we are situated on a local/national or global scale, and in particular if we adopt the Western point of view. I will speak of the Western imaginary and not of the global imaginary, because this book is about tourist practices belonging almost exclusively to Western travelers (European and North American). It would seem that, far from being shared throughout the world, the representations, imaginaries, discourses and practices discussed here are attributable to the West, even though all of this were to spread worldwide. In many ways, Western representations of the favela have more

2 See Cousin (2016) for a discussion of the notion of the imaginary.

advantageous elements, and it seems that the stigma fades with distance. The stigma is stronger among neighbors, and only the "exoticism" that a distant view builds seems to be able to create the conditions for a real interest in discovering the favelas.

However, local and Western representations of the favela are not radically opposed. The Western imaginary of the favelas is made up of extremely diverse, sometimes contradictory elements, and also mixes negative perceptions and idealizations. This contradictory aspect, in any case ambivalent, undoubtedly comes from the way in which it has been constructed by the concomitant discourses, on the one hand, of the Brazilian authorities trying to enhance the reputation of the country or the city and, on the other hand, of all kinds of artists who have progressively made the favela an exotic place. Because it is mobilized at different levels in the construction of tourist practices for which it is a decisive foundation, this imaginary must be commented on here. Throughout the 20th century and up to the present day, numerous artistic or playful works have used the favela as a setting, less often as a subject. By evoking here some of these international productions, we can then understand a part of the Western imaginary on the favela, and thus the formation of a common foundation of expectations, touristic a priori. These images of the favela that are diffused worldwide have, in fact, strongly contributed to shape, not without contradictions, several aspects of the Western imaginary of the favela.

Having become an international brand (Freire-Medeiros 2007), the exotic favela is nonetheless a marginal space, and it is this dimension that is at the heart of many other uses of favelas in film. Narratively, the favela is a codified space located outside of state control; it is the place where the protagonists can escape their pursuers. Already in 1964, the character belonging to Jean-Paul Belmondo hid there with Françoise Dorléac in *L'Homme de Rio*, making the link between the oniric aesthetics of *Orfeu Negro* (1959) and the dynamics of action based on criminal evocation, imitated in the sequel to *City of God* (2002). In fact, it was especially towards the end of the 2000s that American productions made the favela a setting, showing a surprising frequency of narrative uses. Among them, *The Incredible Hulk* (2008) and *Fast and Furious 5* (2011) were very successful worldwide, totaling respectively 263 million and 626 million dollars in revenue[3]. In these films, the narrative and cinematic use of the favela is almost similar. In each case, the favela is the setting for the beginning of the movie: it is easily understood that the protagonists are hiding there and that their hideout is about to be discovered.

The favelas are thus presented as lawless areas, where wanted people can hide. The American heroes, who are eventually found, enjoy the anonymity and seem to

---

3 The American box office, unlike its French counterpart, quantifies the success of films by their receipts and not by the number of admissions. It is therefore easier to talk about revenues when referring to the worldwide success of a film.

appreciate the freedom that this place on the fringe of society offers them. Then – and this is an equally important reason for this choice of setting – once the hideout is revealed, both films stage a huge chase through the favela. These manhunt sequences are extremely alike. In both films, the chase is filmed simultaneously in general views from a helicopter and in closer shots that follow the characters, first through the maze of the favela, and then on the rooftops.

The staging of the favela in the cinema also relies on a vision of the favela that is more exotic, mixing bright colors and rhythmic beats. At the origin of this important aspect of the global diffusion of an idealized vision of the favela, Bianca Freire-Medeiros points to the impact of a film: *City of God* (2002). This Franco–Hispanic–Brazilian film describes the evolution of the eponymous favela (located in the western zone of Rio de Janeiro) in the 1970s and the radicalization of the violence linked to drug trafficking. Through stylized images accompanied by a soundtrack that combines samba, funk and rock, this internationally successful film, which combines extreme violence and great aesthetic concern, has contributed to associating the favela with an image that Freire-Medeiros describes as "glamorous", "cool" and "sexy". *City of God* had a decisive importance on the evolution of the international imaginary towards the favela by conferring to the latter, and while favoring its association with violence, a powerful exoticism, an atmosphere made of warm colors accompanied by bewitching music. Therefore, its impact on the development of tourism was immense, which some critics had already foreseen when the film was released, accusing *City of God* of "promoting hellish tourism"[4] (Bentes 2003). Freire-Medeiros confirms the importance of the film in the process of attracting tourists to the favela. She notes the frequency with which the film is cited by tourists during excursions to the favelas. This is a fact that I have also observed for myself. Many tourists, in fact, asked the guide, during the tour, if the film was shot in Rocinha. But one of the direct and equally important consequences of the commercial success of this film, which has been resolutely oriented towards global distribution since the beginning of its conception, is undoubtedly the fact that it has opened up the possibility of using the favela as an alternative symbol that is particularly commercial.

Thus, simultaneously with or following the success of Fernando Meirelles' film, the favela has appeared, in various parts of the world, at the heart of various commercial strategies that both testify to this "glamorization" of the favela and feed this imaginary. The example mentioned by Freire-Medeiros of the proliferation of restaurants and bars that use the term "favela" is quite significant of this

4 The controversy about the voyeurism approach to the film, then about the practice of tourism in the favelas, refers to that of the film *Slumdog Millionaire*, whose release caused much criticism, especially in India, where it was sometimes called "pornography of poverty", and also enhanced tourism in the slum of Dharavi, in Bombay.

phenomenon. For example, there are club-restaurants called "Favela chic" in Paris, Miami and London. The slogan of the London establishment is particularly telling as to the reasons behind the choice of such a theme, and could also be applied to the film *City of God*: "It's all about exotic flavors, bright colors, and a touch of unusual, in short, a feast for the senses" (ibid.). The term is present in the names of several other restaurants located around the world, including New York, Glasgow, Tokyo and Sydney. In another notable example, the Brazilian owners of the famous Havaianas sandal brand have largely based their marketing strategy on the glamorization of Carioca poverty, and claim that their sandals, now worn worldwide, were originally "typical of Brazilian street children" (Leu 2004).

These artistic productions located in the favelas of the city of Rio de Janeiro have strongly contributed to their inclusion in the Western imaginary of Brazil, while at the same time producing an extremely diverse, even contradictory, discourse on their nature. On the one hand, we have a set of images that play on the marginalization and violence associated with the favelas, to which we can also add the use of the favelas as backdrops for war video games in which we can replay the confrontations between drug traffickers and soldiers. On the other hand, we have a glamorization of the favela, its colors and its sounds, which tends to associate it with the "authentic" Brazil, melting pot of popular music (samba and bossa nova). With this, the distinctive success of the film *City of God* operates a syncretism, maintaining extreme violence and a narrative of the harshness of the favelas, while pursuing an association with Brazilian culture as it is best sold abroad, especially its music. In this, the samba rock of *City of God* operates a modernization of the images, initiated in Marcel Camus' *Orfeu Negro*, which, in 1959, already made an association between bossa nova and a poverty experienced with enthusiasm. These polar opposite images of the violent favela and the glamorous favela, often mixed in popular imaginary, provide an inexhaustible source for the touristification of favelas. Whether they are negative or positive, attractive or repulsive, the main impact consists of the close association they have produced between Brazil, its former capital and its favelas.

The creation of a Western imaginary about the favela through cinema is important for at least two reasons. First, because a large part of the motivation for tourism is to recognize, that is, to come into direct contact with elements that constitute an imaginary that precedes the trip and that often establishes both the list of things that we hope to find, to see with our own eyes, and a repertoire of images and ideas from which to judge the conformity of what we are shown. In *The Tourist Gaze* (1990), John Urry shows this link between the global circulation of images and the circulation of travelers who seek out these images, while creating new ones. The phenomenon is not recent, painting and literature (Bertho-Lavenir 1999) having played this role in the early history of tourism. On the subject of painting, Urry argues that the ways in which the visual interest of sites visited depend on a resemblance to the aesthetics of a painting, and that tourism (in this case, sightseeing) proceeds in its emergence from the "distinction between nature and art dissolved into a circularity" through which "the landscape became a reduplication of the picture that preceded it" (Urry and Larsen 2011, p. 100). Recent examples of tourism development related to cinema, to the places where the films are supposed to take place (the development of international tourism in Montmartre following the success of *Amélie*) or simply to their filming locations (New Zealand for *The Lord of the Rings*, or the city of Dubrovnik following *Game of Thrones*) have proven the potential of cinema to generate desires to visit. Moreover, cinema conditions various associations between places and values. So much so that to visit the places of cinema is not only to find images, but to seek values, emotions and feelings. In this, by what it confers to tourism, cinema accumulates the power of literature and painting. As far as the favela is concerned, this medium has been a central place for the construction of an international brand, which first allowed an indirect experience on the basis of which many desires to visit were developed. Léo, our young tourist who, having just arrived in Rio de Janeiro, rushed to take part in a favela tour, saw *City of God* a few years earlier and remembered it being a shock. He situated there the birth of his attraction, finally satisfied, for Brazil and the favela:

> When I saw *City of God*, it was like a shock for me. Since then, I have listened to a lot of Brazilian music and I have wanted to go too. I also saw the series based on the film, and it's true that when I think of Brazil, I think of the favela (Léo, French tourist visiting Rocinha, June 2012).

Not all the tourists who participate in a favela tour are like Léo, and many of them have not seen any of the films mentioned. The fact that they go to a favela in spite of everything should relativize the importance of this imaginary. It is not the case. Indeed, if the images are not equally shared, they are no less determining, in that they condition the value of a place, or here of a type of place, which is in a way doubled. On the one hand, the favela now exists as an image and, above all, positive values are associated with it. On the other hand, the favela exists as a real place, somewhere in the world, without us knowing if we will ever see it or if it really corresponds to the image we have formed. It is from this double existence, whatever we may say about the correspondences or in coherences between the imaginary and reality, that tourism is born, both as a cultural practice that involves passing from representation to reality and as an industry working to facilitate the visit of places anticipated by the imagination. In other words, it is first as a set of images that have taken a positive place in Western representations of the world that the favela has ceased to be invisible or to be an obscure place on which fears are projected, but not images, and from which we should not approach.

1.2. The favelas of tourism

For this research, I conducted multiple interviews with tourists who had visited a favela. These conversations either took place during the visit, or shortly after a tour I had taken with them, or much later, sometimes years after. In each case, I noticed a striking phenomenon. Most of them forget the name of the favela they have visited, and all of them refer to the term "favela", in the singular, to describe the place of their visit. However, the guides mentioned the names of the favelas they visited, some of them with insistence, especially when they lived there. The favela inhabitants, on the other hand, seemed to refer to their *favelado* status only on specific occasions and when they opposed, or are opposed to, other Cariocas. In most cases, it is towards the favela where they lived that they naturally express their belonging. That the term "favela" becomes, in the post-visit narratives, the only toponym by which tourists refer to the place of their visit proceeds from a denial of the heterogeneity of the spaces included in the term.

There are about 1,000 favelas in Rio de Janeiro. They differ in size, origin, geographical location and more. Some are very small; others are so large and populated that they feed the idea of "cities within cities". In spite of this diversity, the term "favela" has been maintained, and the use of the singular is frequent both for those who refer to it from the outside and for those who live there. What the use of the singular for the favela covers is never stable; the category is constantly evolving and retains a polysemy that is not only inevitable, but constitutive of the very way it evolves over time. Several processes, several competing and even opposing definitions, co-produce a category with blurred outlines, but which paradoxically applies without too much difficulty to a set of urban spaces with varied characteristics. Tourism is one of these processes, and it is mainly located at the level of the category, of the favela in the singular, and a visit to one favela thus seems to be valid for all the others.

1.2.1. *The favela, a relative urban category*

What can be noted from the outset about favelas is that the term is not translated (Wachsberger 2008), both in local and global contexts. Indeed, more general terms or their equivalents, such as slum in English or *bidonville* in French, only appear in particular situations and are used either by those who intend to stigmatize favelas (in eradication discourses, for example) or by those who plan to help them (such as NGOs and tour operators when they need to emphasize the poverty of the place). Outside these specific contexts, the term "favela" always prevails. The favela is thus a particular reality, a specific urban phenomenon in Brazil. But where does the specificity of the situation motivating the permanent use of a term lie? While favelas seem to be largely confused with the slums found throughout the world, the question still arises.

Indeed, if we try to define what a slum is in the broadest possible way, certain universal characteristics seem to emerge: slums are places characterized by overcrowding, lack of access to clean water, poor sanitation and insecurity of the inhabitants regarding the preservation of the enjoyment of their homes. Therefore, many favelas would have the main criteria to be considered as slums. So, what are favelas? The 1950 census identified five main criteria: a settlement of at least 50 people, rustic shack-like dwellings, lack of title deeds, lack of access to public services and lack of paved streets. According to the Pereira Passos Institute (IPP), a favela is "a predominantly residential area occupied by a low-income population and characterized by precarious infrastructure and public services, narrow streets and irregular alignment, lots of irregular shapes and sizes, and an unregistered construction that does not conform to legal patterns". But these definitions reveal at least two major problems when we try to apply them to reality.

First of all, there is a real diversity within what is actually referred to as a favela. Indeed, while we find favelas that have all the characteristics of a slum and correspond to the criteria I have just mentioned, others differ in many respects compared to the official version. Many favelas, like all of the ones from my research, are entirely built in hardened concrete (this applies to the streets as well as the houses) and have access to most public services, contradicting in every aspect the definition used in the 1950 census. Thus, the term "favela" encompasses a wide variety of spaces and housing qualities. While material living conditions cannot constitute an absolute common point between the different spaces that are called favelas, the illegality of their occupation – an argument often used to define them – is no more useful, since many favelas have acquired legal property titles. Moreover, as Rafael Soares Goncalves' research (2010) shows, in some places these titles have existed since the beginning of settlement on the hills, with the owners renting out parts of them to poor people looking for housing in the city. Thus, not all favelas were built by squatters on municipal land.

Moreover, while favelas are all extremely different from each other, the conceptual boundary between the favela and the rest of the city also seems fragile. Through a study conducted on 1991 census data, Licia Valladares and Edmond Préteceille (2000) showed that the separation of urban space on material criteria could not be valid. Comparing what was considered by the IBGE (Brazilian Institute of Geography and Statistics) as a favela with the rest of the metropolis, in terms of "level of urban equipment", "occupation status" and "level of education and income of the head of the household", they showed that the results for favelas were often similar to those of other urban spaces located throughout the city – with the exception of areas occupied by the upper classes.

The favela is therefore not a slum; or rather, some favelas are slums, but not all of them can be classified as such. As we have seen, the term "favela" covers a multitude of different situations. Thus, it can be said that there is "neither homogeneity nor specificity of favelas, nor unity between them, nor even within them for the largest ones" (Préteceille and Valladares 2000). Here, they refer to the differences in living standards in the largest favelas (Rocinha, Vidigal, Complexo do Alemão, etc.). In these favelas, material conditions, land use status and access to public services are not the same for everyone: there is a center–periphery relationship within them. In Rocinha, some homes are made of metal sheets and are built on the edge of the rainforest, on top of the hill where they are particularly vulnerable to possible landslides, while the bottom of the favela has buildings, shops, banks and street names.

Despite all these reasons to abandon it, the fact remains that the category of favela does exist on both a local and a global scale. Indeed, although many problems arise when we try to define what a favela is from an objective point of view, its classification seems uncomplicated outside of this scientific context. But the definition of favelas, and a fortiori of the favela, can only be relative. It is developed from an original opposition between *morro* (hill) and *asfalto* (asphalt). In addition to an important topographical dimension, the contents of this opposition vary. In the past, in the early days of the favelas, it referred to a dichotomy between the countryside and the city, with most of the favelas being rural spaces that were out of step with the urban lifestyle that was developing in Rio de Janeiro. Today, what is included in the categories of *morro* and *asfalto* depends on our position and therefore cannot be summarized objectively. Since the 1980s, the areas where drug trafficking takes place in the favelas has become an important definitional element, associating the *morro* with criminality, which is already present due to the illegal occupation of the land, reinforced by the incessant war between the police and the traffickers. The inhabitants of the *morro* and the *asfalto* perceive and experience the border that separates them from each other.

1.2.2. *Tourist favelas: Rocinha, Santa Marta and Vidigal*

The concept of favela tours is a salient and widely commented phenomenon, and does not concern all favelas. Only a handful of favelas have experienced the organization of group tours, most of them in the South Zone of the city: the richest and most touristy part. During my research, I participated in tours in 15 different favelas: Rocinha, Vidigal, Vila Canoas, Santa Marta, Morro da Babilônia, Chapéu-Mangueira, Pavão-Pavãozinho, Cantagalo, Tavares Bastos, Pereira da Silva, Morro dos Prazeres, Tabajaras, Morro dos Cabritos, Cidade de Deus and Morro da Providência. Thirteen of these are located in or near the South Zone (Morro dos Prazeres and Morro da Providência are located in the Centro), with the exception of Cidade de Deus, which is remote but not widely visited. The fact that there were attempts to develop tourism in these favelas during the golden age of favela tours does not mean that all of them have experienced constant tourism. In fact, three of them stand out, and should be presented in more detail.

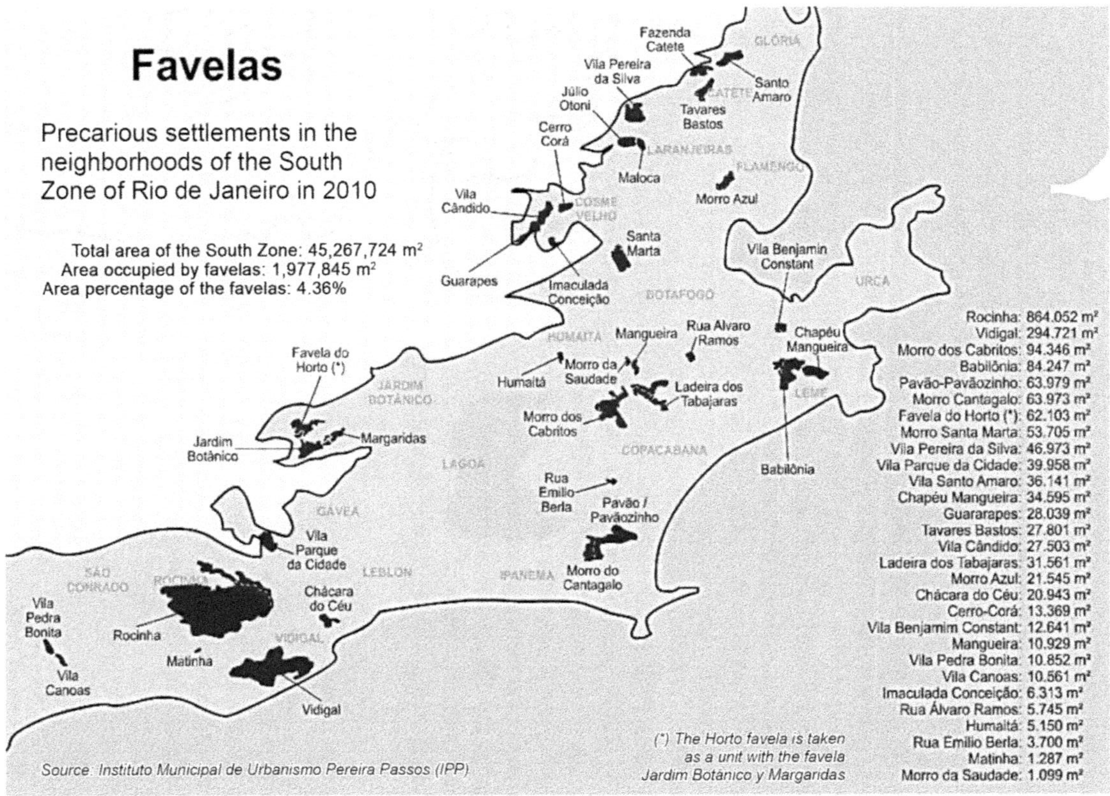

Figure 1.1. *Favelas in the South Zone of Rio de Janeiro (see: https://pt.map-of-rio-de-janeiro.com/favelas-mapas/favelas-zona-sul-do-rio-de-janeiro-mapa-em-pdf). For a color version of this figure, see www.iste.co.uk/apchain/favelatours.zip*

The first one, Rocinha, is considered to be the largest favela in the city and in the country. It is difficult, however, to evaluate with certainty the number of inhabitants of Rocinha, and estimates vary between 62,000 (official data) and 220,000 inhabitants (guides sometimes estimate up to 300,000). Some elements, such as the density of its commercial activity, the presence of national banks and the existence of a middle class in some parts of it, make it an exception among the spaces commonly considered as favelas, of which it is a part despite its official status. Rocinha has a key role in the development of tourist excursions in the favelas. Several hypotheses can be put forward to explain why the Rocinha favela played a pioneering role and remains the most visited favela in Rio de Janeiro. However, what seems to best explain the success of visits to Rocinha is precisely the presence of large tourist agencies (Favela Tour, Jeep Tour, Be a Local) specialized or unspecialized in favela tourism. Indeed, Rocinha remains, by far, the most visited favela by this type of organization. It is also worth looking into the reasons for this preference shown by non-local agencies.

Rocinha has a certain interest for the tourist promoters who can take advantage of its status as the "biggest slum in South America". Far from being anecdotal, this status is an important commercial argument which I have been able to verify on several occasions with tourists. My numerous trips back and forth between France and Rio de Janeiro have given me the opportunity, at random meetings in Paris, to discuss with some of the former visitors of Rocinha. It is certain that the size of the favela gives a certain weight to these visiting accounts. This is how Victor, a 34-year-old Frenchman, presented his experience:

> It is the largest favela in the world. It's amazing, there are houses everywhere. When I travel, I try to go everywhere. If you listen to everyone, they will tell you not to go there, but oh well! I found it crazy to see this huge slum, you can't even imagine (Victor, 34, Paris).

It is interesting to note that when the dimensions of the favela are mentioned, it is quite common to use the term "slum", which can be avoided in other circumstances. This is undoubtedly because the comparative perspective implied by the size of the favela requires the categorization of the space to be broadened. In any case, Rocinha is also the most visited favela, and it is here that local and non-local tourism agencies compete most fiercely. In Rocinha, the organization of favela tours by locals is relatively recent, and it was mostly the agencies run by individuals who do not live there that first developed this tourism. In addition to the argument of the size of the favela, the tour operators first relied on a practical specificity. A road, estrada da Gávea, crosses the favela from the bottom to the top, thus allowing vehicles to cross, which is what the tourists did at first, as we shall see when we talk about the history of the favela tours.

The second favela of particular significance is Vidigal, the second largest in the South Zone of Rio. It is located on the opposite side of the hill from Rocinha, on the Morro Dois Irmãos, in the extension of the neighborhoods of Ipanema and Leblon. Overlooking the Atlantic Ocean, it is visible from the beaches. The relationship between international visitors and Vidigal is quite different from that of Rocinha. Here, tourism is not limited to the favela tours and exists in many different forms. Indeed, Vidigal is a space of less excursionist tourism, which could be described as more experiential, and is distinguished by the development of tourist accommodation: mainly youth hostels located at the top of the favela, thus creating a cosmopolitan space.

Although the word is sometimes used for others, it is likely that Vidigal is the only favela in which we can really speak of a gentrification phenomenon. Hotel speculation is not the only cause of this, and in the last 10 years many foreigners have bought houses in the upper part of Vidigal. Some international celebrities have even bought houses in the area without living there or visiting regularly: the French actor Vincent Cassel, the ex-footballer David Beckham, the American rapper Kanye West, etc. As a result, many inhabitants have left the favela, either, for the owners, by giving in to the pressure to sell their house at a good price, or, for the tenants, being faced with the sudden increase in rents. Vidigal is entering the tourist map in a very different way compared to Rocinha: as a "bohemian favela" that tries to preserve its heritage according to the artists, stylists and foreigners who have recently settled there; and a "chic favela" according to those who criticize the phenomenon. Since 2012, a trendy bar, Bar da Laje, is located at the top and even brings in Cariocas and Brazilians on vacation (which is not the case in Rocinha or other favelas, except for a public that likes *baile funks*, big Friday night parties). The price of drinks is very high, in sharp contrast to the prices found in the rest of the community. On nights when a band plays on the terrace, visitors must pay an additional 50 reais. Vidigal thus stands out, while exposing itself to the critics who denounce its gentrification, by the possibility that a visitor can take their time, staying there for several days, and enjoying a chic and bohemian atmosphere. On its website, Bar da Laje expresses this specificity well:

> [The] tourist potential [of Vidigal] goes far beyond the "favela tour" with workshops, charming bars, parties that are already among the most famous of the city[5].

Finally, a third favela, Santa Marta, forms, with Rocinha and Vidigal, the group of the main tourist favelas. The favela of Santa Marta is located in the district of Botafogo. Smaller than Rocinha and Vidigal, it is inhabited by about 6,000 residents. On December 22, 2008, Santa Marta was chosen by the authorities to be

5 See: https://bardalaje.rio/?lang=en [accessed December 4, 2021].

the first "pacified" favela. Its tourist development makes the favela a special example for this study.

In fact, the state was not content in choosing Santa Marta as the primary target of the pacification program, as well as concentrating its efforts there – the only direct ones – in terms of tourism development. In 2010, the Brazilian Ministry of Tourism, together with the Secretariat of Tourism of the State of Rio de Janeiro, launched the Rio Top Tour program. This program aimed to develop a so-called "community tourism", through the training of Santa Marta residents to welcome foreign tourists. It was a two-year program, at the end of which it was expected that the community would be able to manage tourism independently. The declarations of President Lula, present at the inauguration of the project, were revealing of the hopes attached to the impact of tourism towards the inclusion of the favelas in the city:

> Our generation must make up for lost time so that in a few years our children will no longer call any place a favela and all will be neighborhoods. What has been done here is an example that I believe could be implemented throughout the national territory (Lula, inauguration speech for the Rio Top Tour project[6], August 2010).

Originally, the project was planned to be located in Morro da Providência, the oldest favela in Rio de Janeiro, which overlooks the Central do Brasil train station. Nevertheless, the permanent difficulties encountered by the "pacification" and the fragility of the installation of the Pacifying Police Units (UPP), followed by budget cuts that multiplied in the State of Rio, left Santa Marta as the sole beneficiary of Rio Top Tour. Nevertheless, the project has had a significant impact on the slum.

Installed at the beginning of the project, a small hut still stands today in Corumbá square, where local guides wait for tourists. As a result, Santa Marta is still the only favela where a tourist looking for a guided tour does not have to book online and can instead find a guide spontaneously. On the other hand, the people in charge of the reception at the Rio Top Tour stand also offer to hand out maps, thus leaving tourists to walk alone if they wish. Without securing a monopoly on tourism, local guides have managed to gain a very large share of the tourist market. Through an arrangement between the local guides and the association of Santa Marta residents, tour operators who continue to operate (such as Be a Local or Jeep Tour) must now pay a tax. External independent guides, although they are more easily able to circumvent the tax, are also increasingly avoiding Santa Marta, angry that they sometimes have to report to the association.

---

6 See: http://www.panrotas.com.br/noticia-turismo/politica/exemplo-do-rio-deve-ser-seguido-diz-lula_61034.html [accessed January 4, 2022].

Although smaller, and thus less invested in by the big tourist agencies of the city, the favela of Santa Marta is a tourist attraction which has developed certain successes. First of all, it is located in the district of Botafogo, which counts numerous hotels and youth hostels. The proximity with the tourists makes it a favela that some people visit all the more willingly because they do not have to cross the South Zone to see Rocinha and Vidigal. Then, it happens that the tourists, without knowing the details, get wind of the strong community trend of tourism in Santa Marta. Often hesitant in their choice of a favela tour and concerned about the conformity of the activity with their ethical principles, tourists can therefore be seduced by the perspective of tourism controlled by locals. Finally, the Santa Marta favela is known for having provided the setting for the clip of a song by Michael Jackson, *They Don't Care About Us*, directed by Spike Lee in the 1990s. Following the visit from the American star, a statue was erected on one of the favela's overhanging points. The tourist promotions of the favelas often manage to distinguish themselves by these heritage details, and it is certain that the statue, near which the tourists take their picture, gives Santa Marta an additional attraction.

1.3. The rise of the favela tour, contextual elements

> It [tourism in Rocinha] won't change anything. Now the favela is considered as cool, everyone loves the favela [*sarcastic laugh*]. But after the Olympics, it's certain: tourism is over! You'll see [*lowering his voice*]. And so will they [*laughs*].

Fernando pointed out to me two heavily armed and protected policemen, who passed by a few meters away from us. He had just been introduced to me by Sergio, one of Favela Tour tour guides, while the group wandered among the stalls of the small artisanal market that awaited them at the top of the favela. Fernando had nothing against tourism, and he often stopped to chat with Sergio when the tourists were at the market. But, like other inhabitants of the favela, he felt a certain irony at the sight of tourists, especially during these times. It was 2012 and tourism was in full swing in Rocinha. This was the golden age of favela tours, and while tourism had not disappeared from the favelas, it had never been as important as in this first half of the 2010s. Fernando, with whom I spoke several times afterwards, was in favor of the changes that Rocinha was undergoing at the time, but he only saw them as ephemeral changes, a parenthesis. It was this context, this momentary interruption of the normal course of things in the favela, that he had just exposed. Police officers, inhabitants and tourists were gathered in the same space neither by chance, nor by a strict relation of cause and effect. Tourism in the favelas had existed since the early 1990s; however, it reached its peak, in terms of volume at least, during this time.

Indeed, the favela tours remained marginal for a long time, before gradually imposing themselves as a full-fledged attraction within the tourist offer in Rio de Janeiro. Since the end of the 2000s, favelas such as Rocinha, Santa Marta and Vidigal have been receiving a constant flow of tourists. Tourists are becoming a regular feature, and several groups are crossing paths in the alleys of the favelas. How can we explain the rise of the favela tours and their emergence from the marginal position in which they were born under the critical eye of the world? Of course, the symbolic motivations that I began to mention in relation to cinema have a lot to do with it, and a positive imaginary of the favela is more and more shared. Yet, the boom in favela tours between 2008 and 2016 cannot be explained by the maturation of practices and their symbolic foundations alone. The context that affects the city and the country more generally is an essential element to comment on. As far as the favela is concerned, it is marked by two elements. First, the organization in Rio de Janeiro of sporting and cultural mega-events that pushed the city and its favelas under the spotlights of the international media, and then caused an intensification of the tourist flow. Then, and in direct relation with this agenda, political choices were made that impacted the tourism favelas, in particular around a federal program called "pacification". The launch of this program in 2008 marked the arrival of favela tours into a new era.

1.3.1. *Mega-events and favela tours*

The 2010s were marked in Rio de Janeiro by a succession of events of global dimensions, which intensified the international media coverage of the city and led millions of tourists to visit it. By the end of the 2000s, the city's event agenda was set and was already occupying many public, municipal, federal and national policies: the World Military Games in 2011, the Earth Summit in 2012, World Youth Day in 2013, the Confederations Cup in 2013, and above all the Soccer World Cup in 2014 (with a final in Rio de Janeiro) and the Olympic Games in 2016.

We know the political importance of organizing such events, which, since they have a global dimension, are a theater for the demonstration of the host country's powers. At the beginning of the 2010s, the "Lula miracle" took place and Brazil, in full economic growth, intended to seize these events to assert its new position as a great world power. Since their creation, the World Cup and the Olympic Games have been instruments of soft power used by host nations to demonstrate their strength. But these events also bring in a gigantic financial windfall, which the organizers intend to use for their economic development. Among the most important economic benefits expected are tourism and foreign investment. But no matter how much credit we may give to the idea of these spin-offs, the organization of mega-events always begins with a massive expenditure of public funds. It already costs between $50 million and $100 million to bid on a project, including consulting and event-organization costs. This is nothing compared to the cost of organizing. The 2016 Olympics cost more than $20 billion to organize, of which the city alone contributed $13 billion. In addition to the costs of the Olympics, there were the costs of organizing other events, the World Cup, of course, as well as the costs of all the sports events held in the city starting with the Pan American Games in 2007 (Zimbalist 2017). The public expenditure involved in organizing the events, in a country where many sectors, such as education, health and transport, are still seen as deficient, quickly raised the indignation of part of the population, and protests broke out in 2013 in the run-up to the World Cup and following the increase in the cost of public transport. But the organization of mega-events creates an urgency that governments use to push through costly policies (Gravari-Barbas and Jacquot 2007) because of the exceptional nature of the context (Vainer 2011).

The poorest neighborhoods, especially when they are close to the areas dedicated to hosting the competitions or their spectators, are obviously the first targets. Deemed incompatible with the reception of mega-events, which are intended to display the power of a nation or a city to the world, disadvantaged urban areas are often the subject of policies that range from outright eviction to more complex programs for making urban poverty invisible. Many researchers have worked on these policies in various locations around the world and in relation to various events: in the Philippines for an IMF/World Bank conference (Berner 1996); in China for the 2010 World Expo (Wang et al. 2012); for the 2010 World Cup in South Africa (Cornelissen et al. 2011); and for almost all Olympic Games, including Seoul 1988 (Davis 2007), Sydney (Blunden 2007) and Beijing (Shinn 2009), etc. The history of evictions related to the organization of mega-events also has antecedents within Rio de Janeiro's case itself. Thus, in 1992, eviction initiatives were launched in anticipation of an agenda on the 500 years of the "discovery" of America and the reception of the Earth Summit. More recently, Rafael Soares Gonçalves has pointed out the links between the reclaiming of the favelas by the militias[7] and the government, rather favorable to the expansion of the militia phenomenon, shortly before the organization of the Pan American Games in 2007 (Soares Gonçalves 2010).

Eviction policies were indeed carried out in Rio de Janeiro in the run-up to the mega-events, despite resistance from residents more or less relayed by the media

7 Because of their exclusion from the tourist world, I do not discuss here the issue of favelas controlled by militias. Some information can be found in Zaluar and Conceição (2007) or Araujo de Paula (2011).

(Williamson 2011). It is estimated that 70,000 people were displaced during this period (Broudehoux 2017). In addition, a policy of invisibilization in the favelas also took place. Like eviction, the concealment of favelas from the city's official imagery was a recurring strategy that took various forms. The end of the 2000s saw the creation of several wall projects for a total estimated at R$40 million for its application to 19 favelas, mainly in the South Zone (Borius 2010). It is also interesting to note that the term "favela" does not appear once in the 420-page dossier submitted by Rio de Janeiro when it was preparing its bid to host the Olympic Games. In this file, it is only the "communities" that are mentioned. Equally revealing is the conflict that opposed, in 2011, the city of Rio de Janeiro and the company Google Maps. The latter had provoked anger among some municipal authorities, who criticized it for leaving too much space to the favelas and, above all[8], naming them. The city therefore sent a formal request to the American company to remedy this problem, which was considered harmful to the city's reputation. In 2013, Google granted the city's request: the term "favela" disappeared completely, replaced by the term "*morro*" which, for those unfamiliar with the city, seems to describe them as green spaces, as no street is drawn in the area of the favelas. In 2016, the favelas made a comeback on Google Maps, marking a certain change in the relationship between the company and the municipality, with Google neglecting to mention its participation in the invisibilization of the favelas between 2013 and 2016, and even giving this comeback the appearance of a militant work[9]. A new example, before and during the mega-events, the Maracanã stadium in Rio de Janeiro is always shown from the same angle, thus avoiding showing the Mangueira favela located on the hill behind (Broudehoux 2017). Some advertisements, like the one for Petrobras in 2010, do not hesitate to give an image of the city stripped of its favelas.

However, this has not prevented the favelas from imposing themselves as a major space, integrated into the Western imaginary of the city of Rio de Janeiro, and they are now visited. Total invisibilization is impossible. Tourism in the favelas is therefore a phenomenon that urban policies must take into account. This is one way to understand the role that a new kind of policy, "pacification", has played in controlling the image of the city around mega-events.

---

8 See: https://extra.globo.com/noticias/rio/google-maps-faz-do-rio-um-aglomerado-de-favelas-1655882.html.

9 See: https://www.theguardian.com/sustainable-business/2016/oct/09/invisible-favelas-brazil-rio-maps-erasing-poorer-parts-city.

Figure 1.2. *Petrobras advertising*[10]*. For a color version of this figure, see www.iste.co.uk/apchain/favelatours.zip*

1.3.2. *Pacification*

On November 11, 2011, a few months before my first stay in Rio de Janeiro, it was on French television, on the 8 o'clock news, that I became aware of a major change affecting my future fieldwork. In world news, already turned towards Brazil in anticipation of one of the biggest sports events of the decade, images showing the military operations in Rocinha, the biggest favela of the future Olympic city, were diffused. Without confrontation, the tanks of the Brazilian army entered the favela to bring about what looked like a conquest. Journalists said that the favelas have been taken from the hands of drug traffickers, and a new era was opening up for Rio de Janeiro.

The time of my studies in the favelas of Rio de Janeiro coincided closely with the unfolding of a new phase in the history of the state's treatment of the favelas, considered to be the major problem of the city. From 2012 to 2017, the years of my fieldwork, the development of tourism was inseparable from that of a policy called "pacification". "Pacification" is an initiative supported by the Secretary of Security

---

10 See: https://comitepopulario.wordpress.com/2013/04/15/a-invisibilizacao-da-pobreza-e-dos-pobres-no-rio-olimpico/.

of the State of Rio de Janeiro, José Mariano Beltrame, and the Governor, Sérgio Cabral. It was an initiative that aimed to take back the favelas from the hands of drug traffickers, replacing them with the presence of a community police force. Its deployment took place in two stages. First, the Special Operations Battalion of the Military Police (BOPE) intervened to evict the traffickers. Many of the BOPE's interventions actually ran rather smoothly, as they were planned in advance so that the traffickers could leave the favela on their own. However, the operation was deployed with an impressive military device, whether it was used or not, whose purpose was also to produce a media effect. Each operation and each capture of territory were immediately reported by the national media as well as by the international media, for the most famous favelas. In a second phase, the BOPE was replaced by police regiments of proximity called UPPs (Pacifying Police Units). The police officers selected to serve in these regiments were often young, non-Carioca and relatively better paid than in other services, all of which were intended to prevent the problems of corruption that had historically plagued police actions in the favelas of Rio de Janeiro. The "pacification" was launched in 2008, first in the community of Santa Marta, in the district of Botafogo, before being implemented in a total of 36 favelas until 2013, the year of the last creation of UPPs.

The UPPs were concentrated around the areas planned to host sports events and international tourists. There was no doubt that the horizon of hosting a set of events of global dimensions motivated the launch of pacification, as well as the foregrounding of the issue of security, which had already imposed itself, at least since the beginning of the 1990s, as the main subject of the cariocas political debates. Not only do these events impose a major security challenge, but they also had the obvious effect of putting the city of Rio de Janeiro in the media spotlight for a long time. With ambitions to be recognized as a world power, Brazil had to demonstrate its ability to deal with the "favela problem". In this, the strong media coverage of BOPE's operations was a genuine part of the policy that aimed to send a strong signal, not only to the nation but also to the whole world. In the critical discourse of "pacification", which is very present in the favela tours, the initiative was therefore treated with a certain amount of suspicion from the outset because of the nature of this objective – clearly not improving the quality of life in the favelas, but the smooth running of the major events to come. In reality, there was no one way of perceiving pacification among the inhabitants. As is the case with tourism, opinions diverge and attitudes vary between optimism about a return to calm (in fact short-lived) and lucid pessimism about the real place of the favelas in the city.

However, mutual distrust was established from the very beginning of the pacification process. There were concerns about the power of this community police force, whose officers did not know the inhabitants, at least not at the beginning. On many occasions, the concerns proved to be well founded, and a large number of denunciations of the actions and abuses of the police in the pacified favelas were soon heard. From the outset, the uncertainty about whether the traffickers would leave, whether real or staged, plunged the police in the pacified favelas into a relationship of mistrust with the inhabitants. On the other side, Palloma Menezes (2014) points to the instantaneous appearance of rumors, which reflected the feelings of the inhabitants. Thus, Menezes reports that the installation of surveillance cameras in the Santa Marta favela caused a rumor to spread in the community that the inside of the houses was also being filmed. Similarly, the prohibition of *baile funks* in the early days of pacification led some residents to believe that listening to funk music was also prohibited at home (Menezes 2014). Then, several reports of police violence emerged, some going so far as to report rapes by UPP officers. Even later, serious doubts were raised about the incorruptible nature of the UPPs, and it seems clear that – in some cases, at least – the old way of working resurfaced.

1.3.3. *Tourism and pacification*

The links between pacification and tourism are complex and ambiguous. And it is necessary to separate discourse and practice.

In terms of discourse, pacification was unanimously criticized. It found an already existing framework in a generalized criticism of the state, designated as guilty of the stigmatization and marginalization of the favelas. The state was systematically considered as corrupt and dishonest in the representations of the favela it conveys. From the very beginning of the pacification program, these initiatives were presented by the guides as a policy designed to move the problem away from the places affected by the major sports events, and to allow the staging of government power. The tone was immediately pessimistic, and most of the guides said that the police would leave as soon as the Olympic Games were over, and that the events had proven them right – but the guides were of course far from being the only ones to share this opinion, which was also widespread among the inhabitants. The idea that operations will end with the Olympics is present from the beginning. It was in fact accompanied by a rumor mentioned by Menezes (2014), according to which traffickers left stocks of weapons in the favelas, foreseeing a planned return. Between 2010 and 2016, there was even a pronounced tendency towards nostalgia for the past and the order established in the favelas by the drug traffickers.

In terms of practices, however, we can only observe how the introduction of pacification coincided with a sudden development of favela tours. At the beginning of the 2010s, guides and agencies multiplied, a number of youth hostels opened in the favelas and, during the World Cup and the Olympic Games (as well as during Carnival), the favelas in the South Zone along the ocean functioned as alternative accommodation (in the form of Airbnb rentals, in particular) to supplement a crowded hotel sector. The number of visits was increasing, and several groups were crossing the most visited favelas at the same time, which also led to a diversification of practices and routes taken by tourists. The pacification program was bearing fruit in terms of a decrease in violence: homicides were becoming rarer, and shootings were disappearing from most of the favelas visited by tourists. Tourists, on the other hand, were not aware of pacification, and few had heard of it. On the other hand, a climate of confidence began to surround the favela tours (leaving only ethical concerns surrounding the visits) and they were promoted in all the hotels, including the luxury hotels of Copacabana and Ipanema. The favelas were positioned as tourist sites in their own right and competed with the city's biggest attractions.

While we have seen that, in terms of discourse, pacification was criticized, we cannot say that inhabitants and tourists experienced pacification in the same way. For tourists, pacification, although it could be criticized, was undeniably a process that favored their visit and ensured a feeling of security. For the inhabitants, on the other hand, pacification was gradually becoming a new form of control. Everywhere, the police were in control, and pacification did not change the inhabitants' impression of being criminalized, occupied, etc.

It is worth observing that pacification was presented to tourists as only criticizable because of alleged hypocrisy on the part of the state, and not because of the increasing surveillance practices and the abuses that were progressively increasing. In reality, many guides do not live in the favelas and receive preferential treatment. The local guides, on the other hand, are perhaps less inclined to criticize a process that, in their eyes, has opened up new opportunities for them.

From the tourists' point of view, and despite the symbolic resistance shown by the content of the tourist discourse, pacification has won its bet on two points. First, the invisibilization of the gangs, some of which moved to unvisited areas or were simply disarmed. We no longer see checkpoints, which leads us onto the second point: the affirmation of a freedom of movement without limits. Tourists are experiencing a form of mobility that is free from the shackles of the past, which allows them to penetrate former lawless areas on a massive scale. The other blind spot of tourism is that the counterpart of this freedom of movement is the restriction of inhabitants. Thus, while maintaining a critical symbolism (that of symbolic inversion, not of disempowerment), tourism fully plays the game of pacification and allows the integration of the favelas into a harmonious national narrative: the favelas were present in the imagery of the World Cup and of the Olympic Games. The official FIFA World Cup match jingle featured a young boy playing with a soccer ball from what is clearly a favela overlooking the ocean. In terms of the Olympics, the opening ceremony took place at the Maracanã stadium, in a colorful setting made of the tangle of houses reminiscent of the favelas.

As expected, the period of mega-events gave way to a return of violence. The traffickers were gradually taking up their position in the favelas. And the federal bankruptcy carried, from 2018, a fatal blow to pacification. On October 23, 2017, a Spanish tourist was shot dead during a favela tour. Several countries, including France, officially advised against visiting the favelas. In 2018, there were two shootings per week in Santa Marta (following none between 2008 and 2015) and a homicide every three days in Rocinha. Tourism in the favelas was entering the first phase of decline in its history. Many guides were changing careers, most of the hostels were closing down, outside guides were only going to the favelas in order to gather day to day information, since police and traffickers resumed an internal war.

For tourism, the favelas were thus only momentarily integrated into the city by the freer mobility regime of the tourists, and it accommodated a discourse left free, critical but not outrageously denunciatory. The restrictions to tourist mobility in Rio have always been part of the daily life. This is also the case for the inhabitants of a segregated urban space, where not everyone crosses borders with the same ease. The demonstration, during the 2010s, of the possibilities for tourists to move around freely through the favelas. We could even say that the traffickers accepted the state's deal: to leave the field free for a while. Other parts of the city, the most touristy, were also subject to an intensified police presence, in a show of force.

1.4. History of the favela tours

The specific context of the 2010s thus explains the rise of favela tours, but it does not summarize the whole story. Indeed, the development of tourism in the favelas is an older process, which needs to be revisited. Even though this investigation corresponds to the framework of the context described above, recounting the story of the favela tours allows us to grasp certain mechanisms at work in the development of tourism, whose existence is independent of this context.

1.4.1. *1992–2008: the pioneers of favela tour success*

The favela tours began in the 1990s in the favelas of Rio de Janeiro's South Zone and focused directly on the Rocinha favela, which at the time of this research accounted for the majority of tourist activity. It is likely that tourism in the favelas was born from simultaneous but not concerted initiatives of several individuals. The history of the first favela tours seems to be first of all that of the encounter with a new demand for visiting the favelas. Indeed, several people declare to have taken the first tourists to a favela on the occasion of the Earth Summit, organized in Rio de Janeiro in 1992. This event indeed marked the meeting between international travelers eager to see with their own eyes what a favela looked like and individuals, not yet professional guides, able to organize this visit. At that moment, it seemed that there was a local participation in the organization of excursions in the favelas. The Earth Summit was a symbolic event, but in concrete terms it consisted of a few visits that, although they broke a taboo, did not yet constitute the basis for tourism development. Moreover, it seems that local participation in the tourism of the favelas gradually dissipated during the 1990s, only to return solidly in the 2010s[11]. It was rather from the outside that tourism in the favelas really began. Indeed, the decisive entrepreneurship was that of individuals who did not live in the favela, and who sometimes knew little about it, but who sensed the emergence of a tourist desire for the favela.

The path that led one of them, Marcelo Armstrong, to become involved in the favela tours is significant of these early moments. Armstrong was born in Rio de Janeiro in the late 1950s. At the invitation of a friend, he was hired as a sports teacher at Club Med and traveled from club to club for a few years. His experiences at Club Med, during which he learned to speak four languages, were the source of his idea for excursions to the favelas of his home town:

> For Club Med, I went to Senegal. And there, at times, I was often myself a tourist. And I was very interested in the social side of the country, it was a part of Senegal, a part of African society that I wanted to know about. That's how I started to get the idea (Marcelo, interview, July 2015).

To this idea, which thus interfered with his own practice of tourism, is added the observation on his return of a desire of tourists to learn more about the favelas:

> The first person who told me he wanted to know about a favela was a French friend of mine. At the time, I had no idea about this subject. I was surprised, but it grabbed my attention anyway. My friend went there by himself, he took pictures. He took pictures all over Brazil. He exhibited them, and I saw his photos in Bordeaux. I thought it was great, and then I understood (Marcelo, interview, July 2015).

It was thus in France that Marcelo first understood that a form of tourism could develop in the favelas. According to him, it was only by learning to see the favelas from afar, through the eyes of a European, that such an observation could be made. The way Marcelo found himself in a position to bring tourism innovation to the

---

11 These are general trends, and there are many exceptions in the form of individuals who have been multiplying initiatives for 25 years to make a living from the tourism of their neighborhood.

favelas was therefore largely linked to his international life path. Indeed, it was by experiencing travel himself, and then through his intimate knowledge of travelers in his own city, that he was able to understand the tourism potential of the favelas. This also explains why favela residents, who were far less numerous in the 1990s to experience the pleasures of an international existence, did not play the leading role in opening up the favelas to tourism.

Back in Brazil, Marcelo started by contacting the president of the community association of Vila Canoas, a small favela located near the neighborhood where he lived. The first experiences and the first feedback from the inhabitants convinced him to think bigger. Especially since, a few hundred meters away, there is Rocinha, the largest favela in the South Zone of the city, home to over 300,000 people. There again, Marcelo had to meet with the president of the inhabitants' association and ask him to convince the traffickers who controlled the favela to accept the presence of tourists, which he managed to do. His company, which he named Favela Tour, could then begin to offer tours of Rocinha, whose structure was gradually being built:

> But at the beginning, in Rocinha, I was shy because it's a huge favela and I didn't know anyone. At that time, it was something very rare to see someone with a tourist's face, and I say that even for me [...] It attracted people's eyes, but always in a very friendly way. People would come up and say, "Ah, where are you from? Are you a tourist? Why don't you go there? There's something interesting [...]" And that happened a lot, and I felt more and more comfortable. That's how I started to go around there, little by little.

Marcelo's account of the early days of his activity shows that the inhabitants of Rocinha played a role in structuring the practices, albeit anonymously: the presidents of the associations initiated the negotiations, the traffickers indirectly weighed in on the choice of routes leading to a mutual avoidance, and other inhabitants were themselves informal guides for excursions that, at first, did not know where to go in this large, unknown space.

Although other companies directly followed the creation of Favela Tour, and sometimes were born at the same time, it was Armstrong who most enjoyed the pioneer status. His tours were quickly successful and were reviewed in *Lonely Planet* and *Guide du Routard*, allowing him to counter the criticism that sometimes emerged against favela tours. The favela tours were a commercial success, and Favela Tour became the leader in the field. This success convinced others to enter the tourism business over the next 20 years. Soon, companies such as Jeep Tour, which offered different types of tours in the city, included favela tours in their list of activities. Other profiles, residents or independent guides, began to take tourists when the opportunity arose; some founded tour operators. From Rocinha, the favela-tour phenomenon spread to other favelas – almost exclusively in the South Zone of the city, which concentrates the vast majority of the international tourist flow. However, this diversification of actors, practices and places only reached its peak in the 2010s, thanks to a particular context whose end marked an unprecedented halt for tourism in the favelas.

1.4.2. *2008–2016: the rise of the favela tours*

Although the first favela tours were organized in the early 1990s, the development of the activity was limited during its first two years, and did not intensify significantly until the 2010s, which coincided with the time of this research. This intensification took several forms and had several causes.

In terms of form, it was of course the intensification of the flow of tourists to Rio de Janeiro, and in particular to the favelas, that first characterized the formidable development of the activity during these years. The favelas were gradually being included in the program of many tour operators, which were increasing in number and volume. Most of the city's hotels and youth hostels offered a tour of the favelas to their clients. This intensification of the tourist flow was accompanied by a triple diversification:

– Diversification of tourism actors: tourism in the favelas occupied, in the 2000s, a greater number of people who chose or try to become guides or to profit, in various forms, from the activity. In particular, it was a time that was conducive to the emergence of a greater number of local actors who took over a sector of activity still dominated by outside companies.

– Diversification of places: originally in Rocinha and then, in lesser measures, in Santa Marta and Vidigal, the trend of favela tours gained several other favelas. All in all, for this investigation, I observed excursions and met guides who managed, with more or less success and sustainability, to develop a tourist activity in 15 favelas of the city[12]. However, the phenomenon remains concentrated in the South Zone of the city, with three exceptions, two located in the *centro* (Morro dos Prazeres, Morro da Providência) and one in the West Zone (Cidade de Deus).

– Diversification of practices: at the time of this survey, a wide range of practices could be found in the favelas of Rio de Janeiro, offering variations around the dominant model, which remained that of the excursion. Some of these innovations were linked to the specific characteristics of a particular favela newly included on the tourist map. This was the case of some tours in Morro da Babilônia, where the

12 Rocinha, Vidigal, Vila Canoas, Santa Marta, Morro da Babilônia, Chapéu-Mangueira, Pavão-Pavãozinho, Cantagalo, Tavares Bastos, Pereira da Silva, Morro dos Prazeres, Tabajaras, Morro dos Cabritos, Cidade de Deus and Morro da Providência.

guides linked up with a reforestation association, thus giving rise to an ecotourism tour of a favela. In addition, other favelas were opening up to tourism around very identified practices but not tours. This was especially the case for party tourism, which was gaining ground in favelas such as Castelo das Pedras, famous for its baile funks where some youth hostels sometimes accompanied tourists, or Tavares Bastos where a jazz club had been set up. Finally, the diversification of practices was sometimes simply the result of an effort made by the guides to stand out in favelas already affected by tourist activity. Thus, everyone in Rocinha was looking for their own viewpoints, their own interiors of houses, their own schools to visit, and they organized soccer matches between residents and tourists, samba classes, etc.

These forms of diversification of tourist practices in the favelas were the result of a variety of concomitant factors, and it was necessary to see and consider them at their confluence, at the risk of not being able to clearly affirm which factor was the most important regarding this or that situation.

It is likely that the extraordinary development of the favela tours was also due to the increase in the number of tourists who visited Rio de Janeiro during the 2010s. In fact, the years of favela tour development correspond to the record years for tourism in Brazil, and specifically in Rio de Janeiro. Between 2006 and 2014 (the record year for Rio de Janeiro), there was an increase from 750,000 to 1.57 million tourists. Certainly, this period of constant increase was a good time for the development of new tourist attractions. The favelas undeniably benefitted from this and even became places of accommodation for tourism, especially during the peaks of tourist traffic (around the carnival and sports events), which saw the multiplication of Airbnb-type offers in favelas like Rocinha. These increases in tourist flow were themselves linked to the organization of mega-events in the 2010s, which were responsible for the main peaks in tourist numbers in the city and, more importantly, helped put Rio de Janeiro in the international spotlight.

2

Visiting the Favela

Favela tours vary depending on the guides and the favelas visited. Nevertheless, there is a common form and, without exhausting this (relative) variety of practices, it is possible to make a general description of them beyond the differences in content, duration or route. To this end, I will draw on a description of the excursions organized by Favela Tour, which in many ways represent the initial model of favela visits, a model that has remained virtually unchanged since the early 1990s and which still makes the tour operator the leader in these excursions. From this tour and its particularities, it will then be easier to comment on the differences that were found, during the 2010s, through the proliferation of tour operators and guides, taking tourists to the favelas of Rio de Janeiro.

2.1. A typical excursion with Favela Tour

Favela Tour operates in two favelas, Rocinha and Vila Canoas, and organizes tours there every day, sometimes up to four or five on the same day, for groups of around 10 tourists. Its particularity, but not exclusive to it, is that it has a minibus that takes tourists from the neighborhoods of Copacabana and Ipanema through Rocinha, then to the foot of the small favela of Vila Canoas. The tours cost about 20 euros and are offered in three languages (English, French and Spanish). They last about three hours, but it often takes a full hour to get from the favela to Copacabana and back again. Unlike other tours, especially those organized by locals, the Favela Tour itinerary never varies, it is always made up of the same stages, the discourse itself changing only slightly depending on the guides the tour operator employs. I will now describe this itinerary.

2.1.1. *On the way to the favela*

All Favela Tour excursions start in front of the Copacabana Palace, on the Avenida Atlântica, which runs along the ocean from the main residential area for international tourists visiting Rio de Janeiro. For Marcelo Armstrong, founder of the tours, this is merely a practical choice. It is only because the place is known and easily identifiable, and that it is near the subway and the main places where tourists stay, that the Copacabana Palace is thus used as a meeting point. Similarly, the tourists do not seem to attach any deeper meaning to it. It is thus undoubtedly for me alone, through the dozens of repetitions of these excursions, and because of the nature of my questioning on the favela tours, that the choice of the Copacabana Palace (also imitated by others) assumes an important symbolic depth. Nothing, in fact, better symbolizes the idea I have of the work of these tour operators who, since the end of the 1990s, have been taking tourists to discover the favelas and, even more so, of the way they think about and present their mediation. Indeed, the excursion is constructed as a transgression of borders, where the guide holds the role of a smuggler. In this, the Copacabana Palace, as the starting point of the excursion, constitutes the first stage of a symbolic work based on contrasts. However, it is never directly discussed and, if it was never thought of in a symbolic dimension, the guides, who are busy checking the tourist lists to make sure they have all the clients, have long since stopped paying attention to it, seeing it as a simple meeting point. Tourists, not surprisingly, usually agree a posteriori that the meeting place has a symbolic meaning. However, the idea is never spontaneously evoked, and the sympathy that tourists show towards my perception has, in itself, no real value when it comes to reporting on their experiences and feelings.

However, because the place symbolizes both opulence in Rio de Janeiro and tourism, two things opposed to the favela in the discourses during the excursion, I could not help but consider that it is here, during the rendezvous, where every favela tour really begins. It is probably because the participants do not seem to consider that the tour has begun, until the guide speaks inside the minibus, a few minutes later, that the symbolism of the place does not seem to produce its potential effect. The opposition between a world in which the presence of the tourist is thought of as normal and expected and an extraordinary world to be discovered (hidden, poor, non-touristy) is, however, posed there.

The Copacabana Palace is actually the meeting point for tourists whose hotel is not on the list of places where the guides accept to take them directly. It is therefore the point set for all those who stay outside the neighborhoods of Copacabana (which are picked up on the minibus route before the hotel), Ipanema or Leblon (which are then crossed on the way) and those who, even though they stay in these neighborhoods, are not in large hotels, but in rented apartments or hostels. In addition to travel considerations, the possibility for some people to be picked up directly at their hotel is explained by the fact that it is often the hotels themselves that are responsible for booking the tours, also promising this service. The meeting points are set to 9 a.m. for morning excursions and 1:30 p.m. for afternoon excursions. Favela Tour has been operating in this way almost since the beginning and has kept this meeting point, even though the metro extends further along the Atlantic and to the entrance of Rocinha since 2016. The guides, who wear T-shirts with the company's logo and the official guide card around their necks, wait for the tourists in front of the hotel and take them to the minibus parked in the alley, where they meet the other tourists who have arrived a few minutes earlier or who have checked in at their hotel. As I have said before, it was quite clear to the participants that the tour had not yet started at this point. After greeting the tourists and checking their names on his list, the guide did not address them anymore and asked them to sit in the minibus where they waited for the others, sometimes for a few minutes. If necessary, the guide even dismissed the questions that were asked of him, indicating that he will soon address everyone to answer them. This silence, which often spread among the tourists as well, remained for a long time as the minibus left the Copacabana Palace and took the road towards the favela, making a few stops along the way to pick up the last participants in front of their hotels, the length of which regularly exasperated the tourists.

It was only when the group was complete that the guide, most of the time after having made sure by phone that all of the people who had booked were present, plugged in his microphone and finally turned to the tourists. He always started by introducing himself, then presented the driver as a Rocinha resident. Indeed, although no Favela Tour tour guide is from a favela, the company systematically employs local people to drive the minibuses. The latter plays the role of intermediaries with the community, an important but invisible role for the tourists. The guide then outlined the program for the day. Favela Tour has the particularity of visiting two favelas: Rocinha and Vila Canoas, both located in the São Conrado neighborhood. Although the tour may start in Vila Canoas, especially if the number of simultaneous groups is too large, the tour normally starts in Rocinha. On the road to the favela, in the direction of Gávea, the guide took to the floor to give some background information. The order and level of detail given to this information varies according to the guide's preferences, but there is definitely a set list of points to be covered. In general, everything started with a list of numbers. The favelas, of which there are more than 950 in the city, are presented as the place where a third of the Cariocas live. To introduce Rocinha in more detail, the guide stated that it is the largest favela in Brazil, while acknowledging the impossibility of establishing an exact count of the population, which is somewhere between the official census (40,000) and the figure generally accepted by the inhabitants and often presented as accurate by the guides (300,000). After this statistical introduction, the guide began a long presentation of the problems affecting the favela. Often, this point opened with a question from the guide, who asked the tourists if they had seen the film

City of God. It is not set in Rocinha, but the travelers generally knew of it. While, year after year, the importance of the film and the number of participants in the favela tours claiming to have seen it seem to be constantly decreasing, it still constitutes a reference used by the guides, both to challenge the tourists around a common image and above all to expose their objective of revealing a different reality.

As one of the most common images associated with the favela, drug trafficking is usually one of the first topics discussed. Favela Tour guides explain in detail the history of drug trafficking in the favelas. Drug trafficking dates back to the period of the military dictatorship: at that time, common law prisoners and political prisoners were mixed in the same prisons and shared methods, the former by teaching the means to finance the struggles of the political prisoners, who in turn passed on the vertical organization logics that allowed the consolidation of trafficking. This explains the name of the Comando Vermelho (Red Commando), the most important faction of drug traffickers in the city. After presenting the drug trafficking in Rocinha, the guide spoke about the pacification program. Since my first excursions with Favela Tour in 2012, when pacification was a recent process (especially in Rocinha, which was "pacified" in 2011), the scepticism of the guides towards this program only increased so proudly to a point where, in 2017, and in the face of an obvious failure and a return to the limelight of the drug traffickers, it had become a symbol of the hypocrisy and incompetence of the state. The view of the Favela Tour guides regarding pacification is that the initiative is mainly a media action of the state, which has never had the inhabitants and their living conditions as its main objective, but rather controlling the underprivileged population during the period of organization of the global events that Brazil was to host, especially the city of Rio de Janeiro (World Cup, Olympic Games, World Youth Day). Thus, the guides introduced a critique of the state, which was central throughout the tour. It was interesting to see that the question of the link between pacification and tourism was never raised, as the organization of international tourist visits could, after all, be considered as a mark of the decriminalization of the favela. While this question can be dismissed, and while criticism of pacification by the state is easy, it is above all because Favela Tour can claim a presence that predates the deployment of the pacification program. From this perspective, the guides sometimes did not hesitate to present the presence of military police as a danger, and the pacification program as a disturbance to the order and security of the favela. In any case, what the guides systematically affirmed before arriving in the favela is how it has always been a safe place. To this end, they regularly refer to the existence of a code of honor and tacit laws in force in the favela. Also, this is what Marco expressed when he told the tourists:

> Here, the traffickers impose laws for the community. It is forbidden to steal, rape or kill in the favela. If someone commits a crime, they are

> brought before the head of trafficking and tried. That's why in the favela you are safer than anywhere else in Rio. You are a hundred times more likely to be attacked in Copacabana than in the favela (Marco, Favela Tour guide, July 12, 2015 excursion).

After having thus relativized the dangerous reputation of the favela by presenting *trafficking* as a regulating authority of safety and the state as having upset this established order, the guides generally tackle the idea of poverty often associated with the favela. The following is an excerpt from a typical speech by Beatrice, before the arrival of tourists in Rocinha:

> As you will see, the favela is not a slum. Don't expect to see houses made of wood and metal sheets. Here, people have access to running water, electricity, even the Internet. There are restaurants, a hundred stores, three banks and some of the inhabitants are actually middle class (Beatrice, Favela Tour guide, August 3, 2016 excursion).

As this discourse is developed, which intends to break the tourists' preconceived notions, it is evident that a certain number of them were plunged in a surprise and a deep interrogation. It is undoubtedly by will to take charge of this potential disappointment that the guides work, before the arrival in the favela, to warn the tourists of the gap that exists between the reputation of the favela and its reality.

The guide's speech before the arrival also dwells at length on education, identified as the main problem of the favela. With figures on illiteracy and descriptions of a failing education system, the guides develop the image of a population undereducated by will of the state. Although common to all favela tours, the importance of the theme of education here is closely linked to Marcelo Armstrong, president of Favela Tour. I will return to this point later, when I discuss the training process of tour guides in the company and the structuring political convictions of its founder. In addition to the impact of these political analyses, it is clear that the introduction of the theme of education is closely linked to the fact that Favela Tour has always supported and shown educational projects. Also, talking about the educational problems of Brazil and the favela allows the guide to introduce the visit of an association supported by the tour operator, and to which the presence of tourists contributes. In any case, each theme remains closely associated with a criticism of the state, which is held directly responsible for the growth of the deep inequalities that structure Brazilian society, and more particularly the city of Rio de Janeiro.

The discourse on inequality takes a particular turn by drawing on the characteristics of the urban space of Carioca. In this respect, even though we admit that the transfer by minibus from Copacabana Palace to Rocinha is only dictated by purely practical considerations, the route provides an essential framework for the guide to develop their discourse. Indeed, when arriving in the favela, the minibus only crosses the wealthy neighborhoods of the city: Copacabana, Ipanema, Leblon, then Gávea. It is in Gávea that the Favela Tour guides initiated a discourse on the urban fabric of Carioca and its specificities:

> All over the world, the rich live on top of the cities, above the poor. In Rio, it's the other way around: the poor live on top and the rich on the bottom. In a few minutes, we will arrive in Rocinha, the largest favela in Brazil. There, we cross a neighborhood, *Gávea*, which is one of the most expensive neighborhoods in South America. A few years ago, a UN study took place that determined the HDI (Human Development Index), that is, the standard of living of the neighborhoods of Rio de Janeiro. It was determined that the *Gávea* neighborhood had the same standard of living as Canada. Rocinha has the same level of development as Ghana (Léo, Favela Tour guide, July 9, 2012 excursion).

The minibus ride to Rocinha had indeed the advantage of making very visible the spatial proximity that characterizes the disparate social territoriality of the city of Rio. Also, the guide, who in any case had usually finished introducing the main elements of the context at this point, gave a more precise commentary on the spaces crossed on the way. The estrada da Gávea, a street that the minibus takes and that crosses the favela of Rocinha, connecting the rich neighborhoods of Gávea and São Conrado, is particularly well suited to the discourse of Favela Tour's guides. The group passed by some of the symbols of Carioca's high society. In addition to the dozens of luxurious villas that can be seen on the side of the road, the minibus passed by the PUC (Pontifical Catholic University of Rio de Janeiro) and especially the American high school, in front of which the guides never fail to point out that it is one of the most expensive schools in Brazil, thus repeating their discourse on the inequalities of the Brazilian educational system. Suddenly, the landscape changed. The large ivy-decorated walls hiding the luxurious condominiums and their uniformed doormen gave way to a tangle of two- and three-story concrete houses, garages where workers were busy fixing up a few old cars, a dozen men in safety vests waiting on their motorcycles, electrical cables overlapping in apparent chaos, and a few pieces of trash littered at the side of the road, in front of metal dumpsters full to the brim. The minibus entered Rocinha without any intermediate space, except for a large police car with "UPP" written on it, serving as a real border between the opulent Gávea neighborhood and the favela. A few meters further on the road, the minibus made its first stop in front of a small artisanal market. We were a few steps away from the highest point of the favela.

2.1.2. *Visiting Rocinha*

The small artisanal market that marks the first stop on Favela Tour excursions is entirely for tourists. Most of the tour operators stop here for about 10 minutes. The market has been in existence for about 15 years. It sells amateur art paintings that represent, in a very colorful way, the tangle of houses on the hills of Rio de Janeiro, mostly by the ocean and near either Christ the Redeemer or the Sugarloaf Mountain. Alongside these works, which make up most of the crafts sold, are also many bracelets and bags like those found in most of the stalls along the beaches, as well as other effigies of the favela (postcards, magnets made of salvaged materials, etc.) and T-shirts with inscriptions like "Welcome to Rocinha". Just before leaving the minibus, the guide usually wants to reassure the tourists about the intentions of the tour operator, and insists on not intending to push for a purchase:

> We'll stop at a market first. I'll give you ten minutes to look. Of course you don't have to buy anything, you can just look. If you need anything, for example to negotiate, I'm right here (Léo, Favela Tour guide, July 11, 2012 excursion).

Purchases, in fact, remain relatively rare. During my observations of Favela Tour, I came to the conclusion that no more than one in five people purchase an item at the artisanal market. Moreover, tourists are mostly content with a keychain or magnet, and do not look for larger items. When someone does buy a painting, for example, negotiations are even rarer, and the customer rarely calls on the guide to take advantage of their proposition to help with the negotiation. In addition to the fact, often reported in ethnographies of organized tours, that tourists who have paid for a tour are generally reluctant to initiate new commercial relationships during the excursion; this low tendency to buy can be explained by the fact that the tour begins with the sale of souvenirs imposed by Favela Tour in order to maintain the coherence of the itinerary, which must continue to Vila Canoas in the São Conrado neighborhood, at the foot of Rocinha. The vendors themselves are not insistent and are content with presenting the objects on their stall, explaining the process of their creation. In 2012, during my first visits, this little talk from the vendors of the artisanal market was usually spoken in Portuguese and consecutively translated by the guide into the language of the tourists. In 2017, I was able to observe that a few of these vendors, who had largely remained the same, were now able to deliver a welcome and object presentation lecture in English. In reality, things rarely go any further after the salesmen's introductory speech, so the tourists turn away from the stalls fairly quickly and generally take advantage of the moment to get to know each other, or at least to exchange a few words about the experience of their stay in Rio or in other parts of their trip to Brazil or South America. Then, joining the group, the guide draws the attention of the tourists to the view hidden by the stalls. Sneaking behind, he shows the tourists a landscape that he says is unique in Rio de Janeiro:

> Remember, I told you that in Rio, the favelas were above the rich. Here you have an incredible view, one of the best in the city [pointing to each place]. Here you can see Christ, Sugarloaf, the lagoa, three of the most important sites in Rio (Marcelo, owner and guide of Favela Tour, July 25, 2015 excursion).

It is as much, if not more, to take advantage of a view of the opposite side of the favela as of the market that the tour operators stop at this point of the northern entrance to Rocinha. Moreover, the guides of other agencies who, for one reason or another, do not take the tourists to the market always find another point from which this view is accessible. In this way, favela tours appear to be linked to most of the city's tourist sites, whose overlooking views are an essential part of the tourist attraction. Indeed, most of the sites visited by tourists, including Corcovado and Sugarloaf, offer tourists the opportunity to see from above the combination of the city and the wonderful natural space for which Rio de Janeiro is known. If only for the possibility of offering new views, opening up the favelas, which occupy a large part of the city's hills, to tourism seems logical.

Figure 2.1. *The artisanal market, at the top of Rocinha (photograph by Apchain (2012)). For a color version of this figure, see www.iste.co.uk/apchain/favelatours.zip*

After having taken some photographs, the tourists were invited to go back to the minibus for the continuation of the excursion. The next stage was approximately 300 meters from the artisanal market, still by taking the estrada da Gávea, the only road crossing the favela and accessible to cars. Less than a minute after leaving the market, the group got off the minibus again and was invited to follow the guide into a garage which, through a door at the back, gave access to a terrace overlooking the favela, this time leading onto the other side of the hill. The guide was silent at this precise moment, as if to take the tourists by surprise with a landscape that reveals the quasi-totality of Rocinha and, further, the bars of modern buildings in the wealthy neighborhood of São Conrado that borders the ocean. The "oh my God!", "wow!" or "unbelievable!" comments never failed to leave the guide, always in the background, satisfied with his surprise effect. After allowing the tourists to have a few moments to enjoy the view and take dozens of photographs, the guide approached the edge of the terrace and drew the group's attention. Most of the information he gave at this point had already been formulated earlier. The aim here was to rephrase it, based on the view, whose immensity made it possible to recall that Rocinha is the largest favela in Brazil, that it is located in such a way as to overlook the wealthy neighborhood of São Conrado, etc. The new information concerned the way the favela works. Indeed, the view from the terrace enabled us to see the presence of some of the services, namely water through big blue tanks and televisions with satellite antennas, both well visible on almost every roof of the favela. Then, the speech focused on the intervention of the state. First, the guide drew attention to the public structures visible from the roof: a huge gutter built on the neighboring mountain that prevents once deadly weather, a hospital, a neighborhood of rebuilt buildings (in place of the old bus terminal), a sports center located at the foot of the favela, etc. At the sight of the kites that the children make twirl from the roofs of the favela, the guide sometimes began to explain the old functioning of the traffic organized, when the kites served as an alarm system to warn of the arrival of the police.

Figure 2.2. *Tourist photographing Rocinha from a terrace (photograph: Apchain (2012)). For a color version of this figure, see www.iste.co.uk/apchain/favelatours.zip*

The visit to the terrace, as the one to the market, lasted about 10 minutes. The group then got back into the minibus, which now began a descent towards the bottom of the favela. According to the traffic, the descent could take 5 to 20 minutes, during which the guide did not make a continuous speech, being satisfied the most part of the time to point out to the tourists some elements which either echoed what he said before, or allowed him to illustrate the questions which were put to him. The tourists then took advantage of not being visible to observe the scenes of daily life, interrupting their contemplation only to scrutinize, on the advice of the guide, the manufacture of water containers or the collection of garbage. Almost at the foot of the favela, the minibus stopped at rua da Alegria, the commercial street of the favela that joins estrada da Gávea to autoestrada Lagoa-Barra, connecting the western and southern zones of the city. In this street, the Favela Tour guides intend to surprise tourists with the presence of three banks in the favela. The street enabled us to, more generally, observe the lively commercial activity of the lower part of the favela. In fact, the main purpose of this stage has been, for more than 15 years, to make tourists walk in Rocinha just once. After a short speech given by the guide in the street, during which the tourists seemed relatively embarrassed and fearful because of the important activity of the street and the passage of the buses and motorcycle-taxis which brush the sidewalks, the group was then guided on foot towards the main road that marked the entry of the favela to join the bus.

This stop retains its functions, but also leads, before the end of the excursion, to visit a school support association sponsored by Favela Tour. The company, until 2016, worked with another structure located in Vila Canoas. This one was often empty at the time of the excursions and its importance in the community has been steadily decreasing, causing, in part, the break away from Favela Tour. In Rocinha, the students were still in class when the group entered the premises of the association, located at the end of a small alleyway off rua da Alegria. In order to demonstrate the work of the association supported by the tour operator's activity, the guide never hesitated to interrupt the class, creating a moment that seemed awkward for all. While the guide emphasized the supportive nature of the tourists' indirect participation, the children present, who mostly remain focused on their notebooks, sometimes slipped through the classroom to speak in a very low voice to their teacher, himself reluctant to lend himself to this forced demonstration of his work. For their part, and with a few exceptions, the tourists seemed to share this feeling of discomfort, well aware of disrupting a situation whose importance the guide had nevertheless intensely underlined through a speech on the essential place of education in the potential development of the favela. The guide, without seeming to be really conscious of this shared discomfort, and while he tried to say a few words in English to children who did not hesitate to refuse him this favor, thus often cut short this situation where no interaction managed to be instituted. Leaving the premises, the group went to the entrance of the favela, where a pedestrian bridge designed by Oscar Niemeyer rises and, joining the bus, left Rocinha.

The road that leads from Rocinha to Vila Canoas is relatively short and does not require the guide to make a very long speech. He often liked to talk about the presence of banks in the favela. While no tourist spontaneously asked the question at that moment – which happened very often – the guide asked: "In your opinion, how many times have the banks in Rocinha been robbed?" Faced with this rhetorical question, the tourists seemed, in the few seconds left by the guide, embarrassed. No doubt they saw very clearly through the guide's game whose speech they had been listening to for more than two hours, allowing them to easily understand that the answer was surely zero, but that they were expected, at the risk of breaking the effect sought by their host, to answer naively a gigantic number. Whether they answered zero or 2,000, the guide could not lose anyway:

> Hey no, once. I told you, here you're not allowed to steal, kill or rape. Once, two guys robbed the Banco do Brasil that you saw and the police came. Just a few hours later, the two were found by the traffickers and brought before the chief. We don't know what happened to them, but all the money was returned to the bank the next day (Leo, Favela Tour guide, excursion of August 12, 2015).

Whether authentic or not, this anecdote, which refers directly to a multitude of more or less legendary stories told everywhere about the organization of order by drug traffickers, closes for its tourists the discovery of Rocinha. From the start, the favela of Vila Canoas was presented as foreign to the drug trafficking system. It was also immediately presented as being of a much smaller size. For the group, the small community of Vila Canoas will allow them to get off the minibus and finally enter a favela. The oscillation that became more frequent between the use of the terms "favela" and "community", a fundamental lexical distinction in many contexts, was significant for the differences in the uses of Rocinha and Vila Canoas by the guides:

> The community we are going to visit now is very, very different from Rocinha. Here there has never been a drug problem, even before pacification. Vila Canoas is a small favela that was built by the same workers who built the golf course in São Conrado. We will walk through the alleys there, and you can see what a street is like in a favela in Rio de Janeiro (Luis, Favela Tour guide, July 30, 2016 excursion).

2.1.3. *Stop in Vila Canoas*

Marking a sudden break between beautiful villas and the walls of the favela, the arrival of the minibus in Vila Canoas faithfully repeated the arrival in Rocinha a few hours before. Warned that they would not be getting back on the bus before the final departure, the tourists get off the minibus. Before the change of association in 2016, the tour to Vila Canoas opened with a visit to Para Ti, a school support association similar to the one now visited in Rocinha. As a result of this change, the tour actually was almost finished when the group reached Vila Canoas and followed the guide directly into the small streets. At the entrance of the alleys was a *lanchonete*, a snack bar like the ones we can see everywhere in the city, where the guide introduced the tourists to Romario, the owner who made “the best caipirinhas in Brazil”. For 9 reais, everyone could have a caipirinha prepared and picked up on the way to the departure point, “not like in the kiosks on Copacabana beach”. When I started participating in Favela Tour in 2012, the choice of a caipirinha (which then cost 6 reais) was much more encouraged by the guides. However, in 2015, the caipirinha in Vila Canoas caused Marcelo major problems, when underage American students on a school tour got drunk on it, much to the fury of their chaperones. On the small hand-drawn sign hung above the bar, Romario’s addition of “-18: alcohol is prohibited to minors” commemorates this incident and symbolizes the negotiations with Marcelo, which could have deprived him of a sale, admittedly more common in the afternoon than on the morning tours, but almost guaranteed in case of a passing group.

After this visit to the bar, the group walked down a small alleyway made partially of steps carved out of concrete and arrived at a small plaza with brightly colored graffiti on all the walls and a slide and other children’s games in the middle. During a period that approximately stretched from 2013 to 2016, a woman from the community would systematically set up shop here to sell bags, bracelets and other souvenir magnets, like those found at the Rocinha artisanal market. This shopping interlude, which sold even less than the one at the beginning of the tour and which Marcelo took a dim view of, finding that too many businesses had been grafted onto his excursion, did not exist in 2012 and had disappeared in 2017. In the plaza, the guide generally addressed two points. The first point was about old election campaign posters, which serve as a pretext to evoke the relationship to politics in the favelas, and especially the clientelism that characterized the activity of politicians in these communities. The second point dealt with the improved state of the streets and the fact that they had been given names (of Brazilian states) as a result of the Favela-Bairro program. Hailed by the guides as a necessary program of urban renovation in the favelas, it was sometimes presented as being perceived negatively by the inhabitants, especially because of the additional costs that accompanied the regulation of electricity consumption. While it is true that the history of regulation in the favelas has often consisted of making taxable services that the inhabitants had already managed to organize themselves, the idea of a principled preference for illegality and a refusal to pay among favela inhabitants is a pernicious prejudice that is widespread among a certain part of the Carioca population, who complain of paying twice because of the illegal detour of electricity.

The crossing of the lanes, which allowed the group to close the loop and to return towards the *lanchonete*, was the occasion to approach a last topic. Indeed, the passage in front of a parish of the favela allowed the guide to approach the question of religion and to affirm the central place that Christianity holds in the life of the community. In the alleys, the tourists had the opportunity to photograph the electric cables already seen in Rocinha and to throw some glances towards the interior of the apartments, hoping to thus see what they resembled. After eventually collecting caipirinhas, the group joined the minibus and headed back to Copacabana. On the way, which could sometimes be very long, especially for the afternoon tours, the guide concluded with a final speech centered on the origin of the word *favela*:

> I will tell you why the favelas are called favelas. At the end of the 19th century, at the beginning of the Brazilian Republic, there was a man who lived in the village of Canudos, in the state of Bahia, whose name was Antônio Conselheiro. He began to form a group, very large, of people who were against the Republic and tried to live in community. The government in Rio wanted to suppress the community. They sent the soldiers of the army with the promise of land when they returned. The army went in and quickly wiped out the community of Canudos. Back in Rio, the army asked for the promised land, but the government did not keep its promise. So the army moved to a hill that was in front of the War Department. On this hill, they found a plant that was the same as in their camps in Canudos. This plant was called favela, and so they named the hill Morro da favela, the first favela in Rio de Janeiro (Marcelo, Favela Tour guide, July 27, 2015 excursion).

This version of the favela creation story, which Licia Valladares (2006) refers to as the "favela creation myth" and which is recounted here in an abbreviated form, is systematically mentioned by Favela Tour guides and, with rare exceptions, by all of the guides who lead tourists into Rocinha. Canudos, whose story (Cunha 1985) is itself a founding myth of the Brazilian Republic, obviously does not explain the entire phenomenon of the favelas and has the major disadvantage of obliterating all of the social and racial factors – from the abolition of slavery to a massive rural exodus – that really caused it. In reality, the event in Canudos is the origin of the term "favela", but the name is also applied to other spaces that existed before the war (Abreu 1994).

After finishing, the guide stopped speaking. On the road that followed, which can sometimes last more than 40 minutes, the tourists seem to progressively divert their thoughts from the favela, taking advantage of the moment to take a little nap or contemplate the landscape, and most often to decide with their travel partner(s) on the continuation of their program. They were dropped off on the beaches of Ipanema and Copacabana, most of the time apparently satisfied with their experience and warmly thanking their guide for the opportunity to have known better the inside of a favela.

2.2. Constants and variations of the favela tour model

Most of the tourist excursions in the favelas follow the model of this typical tour proposed by Favela Tour. All of the guides observe each other, and most of those interviewed for this study participate, or have participated, *incognito* in other people's tours, either for inspiration or to distinguish themselves from them. Favela Tour is the model against which other agencies or guides base their tours, often using the essential structure and narrative and visual elements. First, tourism in the favelas is unified by a common clientele since, whatever the tour chosen, the majority of tourists are North American or European. This also means that English is the dominant language on favela tours (followed by French and Spanish). In addition, the duration of an excursion varies relatively little (always half a day maximum, except for special requests from tourists) and the cost is similar (about 20 euros in general). The favela tour offer is globally homogeneous and varies only in minor points (which I will explain below), which betrays the perception, by the professionals of tourism in the favelas, of a homogeneous demand and clientele. Everything works as if the very possibility of tourism in the favelas emerges from a set of elements necessarily reinjected in a transversal way into the entire offer. And, indeed, some of the aspects of the tour that I have described draw the initial model of the favela tours and seem to constitute real invariants to tourism in the favelas.

2.2.1. *The common foundation of favela tours*

First, the favela tours are presented as a border crossing. Tourists never meet directly in the favela, but meet the guide outside. In the case of Favela Tour, as we have just seen, the tourists are grouped together far from the favela and taken there by bus. In other cases, the meeting point is given at the foot of the favela, especially when the guides do not have a vehicle or the favela visited has no road. In either case, the guide begins their presentation before having really taken the tourists beyond the borders of the favela. This is an opportunity, in a purely narrative logic, to give the broad outlines of the context of the history of the favelas and the conditions of their establishment in the city. The proximity, in Rio, between the richest spaces and the favelas creates a narrative of this specific fragmentation of the urban space of Rio. But the crossing of the borders also has a symbolic character that especially confers to the guide a role of smuggler, justifying at the same time their presence and the interest of the visit. The service proposed by any tourist guide who offers the possibility to clients to enter in the favela involves crossing a border that, without intermediary, could not have been crossed. This point, essential to reinforce the guide's importance, is also essential to demarcate the favela from the rest of the city. Throughout the excursion, the favela is presented in contrast to the city.

Access to an overlooking viewpoint is another shared characteristic of the favela tours. In fact, the paradox that consists of their location in the most attractive places of the city – the hills facing the Atlantic Ocean – is one of the most famous peculiarities of the favelas. Therefore, any visit to the favelas must include a stop to a viewing point, which offers tourists a view of the entire site. Most of the points chosen by the guides include both the ocean and the widest possible panorama of the favela visited, which can be seen from the top. In some cases, several steps are even planned for this purpose. This is the case of the Favela Tour in Rocinha, with a view of the Lagoa neighborhood and Corcovado on one side and the favela down to the ocean on the other side. For the access to these views, a trade has even developed between the guides and the inhabitants who rent the passage of the groups by their terrace. It should be said here that the imperative of the view is not specific to the favela tours. Guanabara Bay and the way the city is wedged between the ocean and the rainforest are the strongest symbols of the western imaginary of the city, and therefore of tourism. In Rio, the most famous tourist sites (Christ the Redeemer, Tijuca Peak, Sugarloaf, not to mention helicopter tours and paragliding) attract tourists to the heights of the city, and the reward is an all-encompassing view of the city and its natural setting.

Another common point is that the favela tours systematically include the visit of an association, most often a school or a cultural structure, which they contribute to financing. The purpose of this step is obviously linked to the ethical questions that weigh on the favela tours and that, in many cases, focus on the question of the distribution of the profits of tourism. Claiming a contribution to the solidarity economy of the favelas is an important point for all guides, whether or not they are favela inhabitants, but it is a vital point for the communication of non-local agencies. Their contributions are scrutinized, they are important for their clients and the guides constantly justify the contribution to the favela that their tour represents, showing that it exceeds the simple symbolic profit that these excursions in a stigmatized territory are supposed to generate by the simple presence of the tourists. Most of the guides choose a school support association, because of the place that the question of education occupies in their discourse on the favela and its problems. Others, provided they live in the favela, finance associations that they have themselves created and run, and which, ranging from a DJ school to an English school, can be chosen more freely.

Among the topics always mentioned in all the favela tours, the function of public services holds a central place. By walking or simply standing in one of its streets, tourists are led to understand the functioning of daily life in the favela. The tangles of cables in the alleys, the water towers on the roofs and the garbage management systems are subjects discussed by the guides when the group faces them. From a symbolic point of view, this is an opportunity to break a preconceived notion that public services do not reach the favela, while giving credit to the ingenuity and resourcefulness of the inhabitants who obtained them, either illegally, for example by pulling cables from the *asfalto*, or as a result of political struggles that forced the state to expand its scope of action.

Finally, the favela tours give a place to the inhabitants while restricting them in most cases to two possible roles, that of craftsman or performer. The favela tours include various stages whose aim is to open up a market interface between tourists and inhabitants. These stages take place mostly at fixed points, where the merchants wait for tourists who are their only customers. The sale of small objects and pictures bearing the effigy of the favelas is the dominant form of this trade. The most visited favelas all have one or more points of this type, either in the form of a fixed market (as at the top of Rocinha or next to the statue of Michael Jackson in Santa Marta) or in a mobile way. The favela tours have also given some inhabitants the opportunity to become performers, and in the most visited favelas there are a number of them waiting for tourists to offer them a capoeira show or a percussion concert. Very rare in the early days of the favela tours, these performances seem to have multiplied with the development of a more communicable tourism and concretely with an increased possibility for the inhabitants to form partnerships with local guides. They are paid by the hat with some success, which increases as the performers succeed in involving the tourists and including them in the show.

It is around these systematic steps that the guides develop the main themes of their discourse (the functioning of the favela, the ingenuity and resourcefulness of the inhabitants, the stigmatization against which we must fight) and try to meet the main expectations of the tourists (safe, by accompanying them beyond a normally impassable border; ethical, by showing the redistribution of profits from tourism; visual; informative). All of this constitutes the common base of all the favela tours, but the excursions vary on other points.

2.2.2. *Main variations of the favela tour model*

The tour described in sections 2.1.2 and 2.1.3, and referred to as the initial model, correspond to the form of tourism that has prevailed for many years. However, the explosion of favela tours in the 2010s has been accompanied by a diversification of supply and the appearance of new practices, added by new guides and tour operators to the initial model. At a practical level, many companies have tried to innovate by expanding the spectrum of languages offered for favela tours. Between 2012 and 2016, favela tours were offered in Hebrew and Dutch, but still not in Portuguese, unless specifically requested. As I have already mentioned, it is also through the tourist opening of new favelas that diversification takes place and, in this case too, the way in which guides position these new spaces in tourism often consists of structuring their excursions in relation to the favela tours as they are practiced in Rocinha. For example, Morro da Babilônia stands out because of its focus on ecological issues. The visit to the favela is coupled with an excursion to the adjacent forest, to meet the inhabitants who have founded a reforestation association sponsored by the visit. Another example is the attempt at tourism in the Morro da Providência favela, which can claim to be the first in Rio de Janeiro and which also included a church from the end of the 19th century, making it one of the oldest monuments located in a favela – and even a curiosity on the scale of the city, insofar as Rio de Janeiro has relatively poorly preserved its ancient built heritage. For a variety of reasons, among them its location (in the center of the city) and significant problems of violence, the idea of heritage tourism has not allowed for the sustainable development of tourism in Providência, and tourist incursions have often been limited to taking the cable car that links the favela to the Central do Brasil train station. Other guides in other favelas have tried, with varying degrees of success, to take advantage of the development of favela tours by playing on local specificities (they could also have mentioned Morro dos Prazeres, which offers an artistic tour along the graffiti covering the walls of the city, or, to a lesser extent, since its tourism is almost as old as in Rocinha, Dona Marta, which represents community tourism), without, however, competing with Rocinha in the long term. Rocinha remains the place where the flow of tourists is the most significant and where the number of tour operators is the greatest, while presenting globally the same practical (price, duration, etc.), formal (types of stages, mediation by the figure of the guide) and symbolic characteristics. In addition, tourism in Rocinha diversified momentarily during the period of its development, and it is therefore appropriate to evaluate these variations of the initial model regardless of local differences.

First, the emergence of more local guides has brought with it some changes in the model. The local guides, in fact, intend to vary the model of the non-local agencies and use the local character as a commercial argument, putting forward the idea of a less superficial visit and opening the possibility of a more lively favela experience.

From a formal point of view, the emergence of tourism organized by guides living in the favela has not led to a total overhaul of the model of the outside agencies. It differs mainly in the absence of vehicles used for the tours, which are therefore entirely pedestrian, but walking through the favela is not unique to the excursions of local guides and other agencies do the same. What represents a real innovation is the possibility of showing tourists the inside of a house in the favela: the guide's house, most of the time. Getting inside is a symbolic issue in at least two respects.

First, the interiors shown to tourists contrast with the preconceived notion that favelas are slums. There are many decorations and high-tech equipment (televisions, game consoles, etc.). Second, taking tourists inside a house, especially the guide's house, is an effective way to make the case for a tour with a local, as outside agencies cannot afford what would be considered an intrusion. However, some locals open their houses to guides, even from outside, for a fixed fee, or more rarely in the hope of a contribution from tourists, sometimes in exchange for the sale of homemade crafts. I will soon move onto a case study. From the point of view of tourists, the possibility of entering a house is unanimously perceived as an added value to a tour that can often seem superficial.

Local guides often promise the possibility of a meeting with the inhabitants of the favela. However, this meeting never really takes place, as tourists and inhabitants often do not have the means (mainly linguistic) to communicate without the intermediary of the guide. As is the case with the tours of outside agencies, the inclusion of the inhabitants during the visits is most often limited to the interactions they have with the guides, sometimes translated by the latter according to their desire. However, these interactions are more numerous in the case of local guides and give the perception of a tourism that is more integrated into the daily life of the favelas. It is not uncommon for tourists to appreciate this fact:

> You can see that he [the guide] knows everyone. For me, that's important. You feel less like an intruder (Jack, American tourist).

And:

> We feel that the inhabitants are not against tourism and, in my opinion, it is especially because we are with one of them. You can see it, they don't stop talking together, laughing. It's great (Patrick, Belgian tourist).

Another example, more nuanced this time:

> Sometimes it's a little annoying: Sergio stops every two minutes and we stand there, waiting for him to finish his discussion. But hey, that's the game, and it's a good sign that he's integrated into the community (Rolland, French tourist).

The relationships that the guide builds with the favela residents are important for the tour, but do not always replace a direct meeting with them. It even seems that, in the case of tours organized by outside agencies, which also include this kind of interaction through the relationships that the guide establishes with the inhabitants during the tours, they are absolutely no substitute for the meeting, which remains one of the most frequent absences raised by the tourists. The difference between local and non-local tours is undoubtedly to be found elsewhere, and it appears that if local guides are more successful in reducing the feeling of superficiality in the interaction with the inhabitants, it is above all because they themselves embody the inhabitant. Indeed, it is a considerable advantage for local guides to be able to anchor in their own experience the story of the functioning of the favela, its daily life and its history.

In addition, the emergence of local tourism is also tantamount to an experiential turn, which aims to increase the participation of both locals and tourists. By organizing a meeting around an activity, some guides choose to depart from the model created by Favela Tour, whose superficiality they point out. These activities are varied: capoeira or samba classes; soccer games between tourists and children from the favela, etc. In addition, participation in *baile funks*, typical favela cariocas parties, has also been the subject of organized tours, although often in the favelas where favela tours are not given, notably because pacification has for a long time proscribed the *baile funks*. The diversification of tourist practices in the favelas was also based on a call for greater immersion of tourists in the favela. Here, the Favela Tour excursions act as a negative model, denounced for its superficiality. All the elements described above, from activities to the valorization of local specificities, work together to reinforce a greater sense of immersion. The duration of the tour also participates in this strategy, and some guides offer longer tours (a whole day, for example), contrasting with the speed of classic tours, where tourists often spend as much time in the traffic jams as in the favela.

3

Advantages of Tourist Mediation: The Guides of Rocinha

This chapter is devoted to the strategies deployed by different individuals to gain the roles of intermediaries between tourists and the favelas, and thus mainly concerns the tour guides. The question of intermediation, its access and its conversion into other forms of profit (symbolic and/or economic), is essential because it determines the type of tourism, the content of the visits and the narratives given to tourists in the favelas. Furthermore, it partially determines how tourism can create new political arenas for the favela. The individuals who become intermediaries between tourists and favelas fall primarily into two categories. First, tour guides from outside the favela have been the main pioneers of what is therefore not community tourism, but excursion tourism conducted entirely from outside of the favela. Then, favela inhabitants have themselves managed to ensure intermediary tourism, thus modifying parts of the discourse and practices, while taking up without really upsetting the model built by the external guides. But, in reality, the distinction between external and local guides hides four different profiles that correspond to as many goals and strategies.

Given the nature of their roles, the guides can be considered as cultural brokers, that is, individuals who benefit from their position as intermediaries. These cultural brokers are defined primarily by the specificity of their skills, that is, by their understanding of the socio-cultural and linguistic codes of the two groups they mediate between. Tourist guides are one of the most obvious examples of professional cultural brokers. The profits they make from intercultural mediation are varied, and are not limited to financial income. Becoming a cultural broker is often a social repositioning strategy deployed on both local and international levels. In other words, these individuals convert their professional roles as intermediators into a revaluation of their social status. In his book *Anthropology and Development: Understanding Contemporary Social Change*, Jean-Pierre Olivier de Sardan (1995) develops a typology of actors he calls "development brokers". In his perspective, "development brokers" are "social actors implanted in a local arena who serve as intermediaries for the draining off (in the direction of the social space corresponding to this arena) of those external resources commonly referred to as 'development aid'" (Olivier de Sardan 2008, p. 173). Here, it is obvious that the concept has been approached by shifting it to a sphere significantly different from the one he studied. Indeed, the structuring of a tourist activity in the favelas of Rio de Janeiro does not correspond to the model discussed in the work of this author, that is, that of "development projects" carried by international institutions. It is nevertheless true that, in the case of tourism, the guides are intermediaries between Western travelers and a local population whose cultural identity they are responsible for transmitting. It is even through the appearance of these guides as cultural brokers that we see one of the most interesting impacts of the development of tourism in the favela. Indeed, the emergence of these new actors constitutes a break from the old political structures, and we do not find the same individuals in the position of intermediaries between the favela and the outside world. The typology that Olivier de Sardan proposes about these "development brokers" is useful for understanding this point better, and allows us to better situate the guides, in their diversity, in relation to their new roles in the particular social arena of the favela.

Olivier de Sardan believes that the different "development brokers" can be distinguished by four main cases. The first case commented on by the author is that "the broker is on the outside of the local political arena, and would like to get in" (Olivier de Sardan 2008, p. 176). In this case, the strategies of intermediarity are combined with the implementation of strategies to establish themselves in the local arena. Second, "the broker is a powerless or low-status or marginalized actor in the local political arena" (Olivier de Sardan 2008, p. 177). Here, the strategies for capturing intermediation and local social ascension are more than correlated, and the whole purpose of acquiring intermediary status is to make it possible to benefit locally. In another situation, "the broker is a high-status actor or occupies a central position in the local political arena" (Olivier de Sardan 2008, p. 177). As in the previous case, the purpose of intermediarity is understood locally but as a strategy of consolidating power (which, moreover, is often made necessary by the very appearance of the change that intermediation requires), and no longer as an acquisition strategy. Finally, in the fourth case, "the broker wants to get out of the local political arena" (Olivier de Sardan 2008, p. 177), and the advantage of acquiring an intermediary function is thus in the hope of an external ascension.

3.1. The guides, a variety of profiles and issues

In order to better understand the different types of guides who take on the responsibility of mediating between tourists and the favela, it is essential to describe the backgrounds and positions of those who embody them. The following information will not exhaust the range of profiles that make up this group of key players in the promotion of tourism of the favelas. It will, however, give a sense of the similarities and differences that these professional mediators display and that, through their activity, fulfill different objectives and thus make the favela tours take on varied forms and contents. It will also keep in its center a distinction between local and external, while showing the porosity of this opposition.

3.1.1. *Marcelo and Roberto, the external companies and the distanced gaze*

The vast majority of non-local companies are structures that employ professional guides who have completed the legal training required to be an official guide in Brazil. They hire them on a permanent basis, usually refusing to allow the guide to work simultaneously in other structures. It may happen that the guides are authorized to carry out an independent activity in parallel, but in no case is it tolerated by the agencies for them to lead tourists independently through the favelas visited by their tour operator. Furthermore, not all companies that have a significant activity in the favelas are dedicated to them exclusively. Also, some companies, such as Jeep Tour and Be a Local, are classic tourist agencies that operate throughout the city and have gradually added the favelas to their offer. It is this plurality of activities of the non-local companies that makes them more sustainable, especially because they can stop visiting the favelas if necessary without completely interrupting their activity, and also because they have a larger customer base. In fact, some of them stopped visiting favelas after 2017. Their guides are chosen according to classic criteria, which apply to other tourist activities in the city. In addition to holding an official guide card, guides are expected to have a good command of at least two languages (English and one another) and mediation techniques. Apprentice guides are trained by following several tours of seasoned guides, which has the effect of homogenizing discourses. In some cases, such as Favela Tour, the apprentices are given selective press clippings to form a homogeneous and coherent discourse on the favela.

While talking about the history of favela tours, I mentioned Marcelo, founder of Favela Tour, and his biographical background. Today, Marcelo is the head of the biggest tour operator that organizes daily visits to the favelas. While, for other reasons and in other favelas, other individuals seem to have simultaneously created favela tours, Marcelo has demonstrated a level of innovation that has allowed him to reach his current position. This innovation, we have seen, was largely inspired by his background and experience of European tourism in Africa, acquired during his time at Club Med. Indeed, it is by working with tourists and – equally important, if not more so – by experiencing tourism himself that he was able to grasp the potential of tourists' interest in the favelas. He understands, especially from his European friends, that there is an interest in the other side of the postcard (the picture perfect view of the place), that tourism is not entirely trivial and that those who travel do sometimes seek to access what he calls the "social side" of the country. While working in a Club Med in Senegal, he himself was attracted to this aspect. In addition to this experience of tourism in Africa, it was the acquaintance of travelers in Rio de Janeiro, his home city, that played a decisive role. Thus, he first had a French friend, a photographer, who went to a favela. It was this event that definitively convinced him of the potential of the tourist excursions organized in the favelas, of which he was not really aware at the time. In this, we can consider that it was the double experience that Marcelo had, as a tourist and in contact with tourists, abroad and at home, of the importance of what seems to be located outside the tourist bubble (of which the Club Med is a typical example) that led him to become aware of the potential attraction of favela tours.

The moment when Marcelo decided to start developing tourist excursions in Rocinha and Vila Canoas is decisive in many ways, and the excursions will always bear the mark of his experience. It shows that an experience of leisure mobility acquired through his personal practice and his professional path at an international level is the key element to understand how Marcelo was able to grasp the interest in favela tours. We will see through other examples that the experience of an international trajectory is a sine qua non condition to be able to pose as an intermediary between the tourists and the favela. Marcelo, by his experience, understood the importance of what lies beyond the postcard for contemporary tourists. And making it a lucrative activity does not for him come with ethical considerations, having himself shared this attraction in other countries. In a way, we can consider that the fact that he did not perceive the attraction of the favelas earlier, and that he only understood this in contact with tourism on other continents, sustainably structured his favela tours. Indeed, Marcelo was persuaded that the decisive element to succeed in mediating between tourists and favelas was through the understanding of tourist aspirations, and thus it was a priority to understand their point of view. As a result, Marcelo recruited individuals with a similar background and who were able to understand what he believed tourists were looking for when visiting the favela:

> I already had local guides, it did not work. With the type of approach I wanted to take, unfortunately, I did not find in Rocinha a capable guide. This is because I needed a guide who has a minimum understanding of Brazilian society: they needed to know a little more about the world, about the kind of desires a Frenchman has when visiting Brazil, how it is for an American to be there… A guide who

> has been to the United States, to Europe, who knows a little bit more than Rocinha, always Rocinha, Rocinha yet again, it's too limited all of that, and it doesn't appeal to me (Marcelo, interview, July 2015).

The guides' discourses during Favela Tour excursions reflect Marcelo's convictions and his vision of the tour as being above all a means of expressing a critical point of view on Brazilian society. Of course, the convictions expressed by the guides are not entirely the same as those of the founder of Favela Tour, and sometimes what is said is slightly out of line with the neutrality of information desired by Marcelo.

The training of the guides, who are all official guides at Favela Tour, is very simple. The new member of the team, recruited for their understanding of tourist expectations, receives from Marcelo a folder containing a few press or university articles that provide them with information they will have to include in their speech during the excursions. This file of articles, which treat, for example, the origins of the organization of the traffic in the favelas, represents the repertoire of Marcelo's ideas on the favela and on what the guide can say, teach or symbolize about Brazilian society and its problems. In addition to learning about these documents, the future guide must take at least five tours led by five different people within Favela Tour. Marcelo's message and point of view are therefore transmitted through the recruitment and training of guides. However, the guide's actual work leaves them a lot of room to deviate from the message, to add some matters that are important to them or to insist more or less on a specific element. One of Marcelo's points of honor lies in the primacy given to what he calls the "great context", and in the setting aside of a point of view too narrowly focused on a favela in its specificity. This point is respected by the guides almost without exception. On the other hand, the content of the discourse can change significantly, creating some problems from Marcelo's point of view, as shown by the example of his argument with Roberto, a guide employed by Favela Tour.

Roberto is a guide in his 50s. *Carioca da gema*[1], as he likes to remind everyone, he worked for a long time in cinema, a profession that allowed him to travel around the world. After living in London for 10 years, he returned to Rio and obtained a guide's license. In addition to his independent activities, he worked regularly for Favela Tour. Roberto and Marcelo share a passionate desire to show international tourists the problems of their society. Like Marcelo, Roberto has no ties to the favela, and sees the visit as a simple opportunity to see Rio from outside of the postcard. For him, not ignoring the favela is like fulfilling a duty. It is in this sense that Marcelo and Roberto see the logic and justification of their work in the favela, that is, as a somewhat militant activity in which their political opinions shine through.

1 Expression used by the Cariocas to mean born and raised in Rio de Janeiro.

It is obviously this sensitivity, "rather left-wing" according to Marcelo, that had convinced him to hire Roberto and entrust him with groups visiting the favelas. It is, moreover, very difficult to make one or the other speak about the favela or tourism, so much each discussion is brought back to a more general debate on Brazil.

Roberto's tours are not very different from those of the other Favela Tour guides. In fact, in terms of the structure, they are identical. The variation brought about by Roberto's personality is only noticeable at brief moments, when he adds a clearly anti-capitalist view to his criticism of the state. Present, but in a less frank way in many elements of common discourse, this view is especially strong when coming from Roberto, who resorts to anti-capitalist and alter-globalist references, notably Noam Chomsky or Naomi Klein. In the few years I knew them, I observed that the relationship between Roberto and Marcelo, which ended with Roberto's dismissal, was quite relevant to Marcelo's control and to the point of view he wants his guides to adopt. Marcelo criticized Roberto for sharing a too "ideological" discourse, which he found to be unsuitable for the message he wanted to transmit through his tours and especially incompatible with the interests of the tourists. Marcelo's distrust in Roberto's discourse continued to grow, until his dismissal in 2016, as the Brazilian political crisis became more evident. However, in 2015, Marcelo was already concerned about Roberto's discourse:

> With Roberto, there were problems, I must say. He was very good. He was a friend of mine before becoming part of the team, but I had to talk to him because he was very (too) ideologically biased. He was very PT (Workers' Party). It bothered me. I had a meeting with the team to talk to them about this. I was a PT too, I had voted for Lula before. But after two, three years, this was enough for me to say: "Oh, what is this? This is not at all what I imagined, it's a lie". But I have nothing against the left. I'm a leftist. But the extreme, stupid, idiotic, tendentious left… In my opinion, this is a lie, he is a liar, he wants to hide the reality to show a left that is convenient for him. We had a meeting to [only] talk about this. And Roberto, he was very left, too left (Marcelo, interview, June 2016).

In 2015, while the operation Lava Jato[2] revealed more and more cases of corruption within the Brazilian Labor Party, Marcelo was keen to incriminate the political class for the problems he denounced in his tours. While he allowed, in a general way, a certain freedom to his guides during the excursions, the national

2 Operation of national scope led by Judge Sergio Moro, leading to the revelation of a multitude of corruption and money laundering cases. The scandals affecting Petrobras and the Brazilian PT, revealed as early as 2014, have largely contributed to pushing Brazil into an economic and political crisis.

situation seemed to plunge him into a deep disagreement with Roberto, whose obvious support for the PT appeared as unacceptable to him:

> You see something in Brazil that I have rarely seen elsewhere. It's a little bit similar in Venezuela. People who are really enraged and people who still defend are a minority, an ideology against a majority of people who are more and more enraged. And the discourse of these people, including Roberto, is [taking a condescending tone]: "No, don't you see that this is a type of conspiracy of the media, of the bourgeoisie and international interests, of the superpowers against the strength of Brazil! Don't you understand that?" Ah no, that's silly, it surprises me. And Roberto, he is very cool. He has made a movie, he is a filmmaker, he directed Anthony Quinn. He's cool, but how can he? How can an ideology be so limited? It's like a religion, it's a silly kind of fanaticism. Defending the indefensible (Marcelo, interview, June 2016).

For Marcelo, Roberto's point of view, which he considered as naive, went beyond the limits of the differences he was willing to tolerate with his guides. Marcelo's choices were not limited to commercial strategies, but were strongly influenced by the informative mission he felt invested in. In a way, it is clear that Roberto's tours challenged his control over the dimension of favela tours, which he wanted to be educational and not political. In 2017, he told me about Roberto's dismissal:

> Marcelo: He crossed the line, so I told him:
>
> "Go…
>
> – Oh, but no, I have to work, blah blah blah.
>
> – Too bad, go. You embarrass me as a citizen, because you are intelligent, but your discourse does not represent the truth".
>
> Because the truth is the truth of the day, not the truth that you've built into your mentality. So I had to fire him.
>
> Me: What was he saying?
>
> Marcelo: No, I saw and heard. I had the proof. I listened. He changed historical facts to favor his very iconic vision of Lula, who has five or six trials in the Brazilian Superior Court against him, and who soon will be, I hope, in prison […]. A few months ago, despite the crisis,

> the disaster, all the facts. I will not pay a guide who, first of all, does not tell the truth. Because his discourse is not the truth, it is his truth, not the truth of Brazil. It is this ideological discourse he protects. But I cannot pay a guide who, first of all, does not tell the truth.
>
> Me: And how did it go?
>
> Marcelo: Two tourists told me, "One of the guides made a communist, anti-American speech". And that I can't accept. So I had to hear it for myself. I called my [minibus] driver:
>
> "Leave your phone on.
>
> – OK".
>
> I hear, "Blah, blah, blah". I said, "Okay, no way", and I fired him. It's a shame, but that's the reality. He can't hide his ideological outlook (Marcelo, interview, June 2016).

The story of this quarrel between two friends, and then of the dismissal of an employee by his boss, says a lot about the way Marcelo conceives his professional activity. For him, it is neither by exoticism nor by voyeurism that tourists go to discover the favelas. More precisely, it is not what he sells. While he admits that a certain number of his customers have a more trivial relationship with the favela, he prefers to concentrate on the educational aspect of the visit. It is indeed a political (but not subversive) function that he believes to fulfill through his work as a guide and director of a tour operator, as well as through the construction of excursions conceived as a critical look on Brazilian society. This is, for Marcelo and many of his competitors, the primary vocation of favela tours: not a deep dive as such into the favela, but an enhancement of what a visit can bring in order to show other aspects of the Brazilian society usually inaccessible to travelers. Generally speaking, tour operators outside the favela share this point of view and give their excursions in this same dimension.

By accepting, as I did, Marcelo's position as a pioneer in the development of tourism in the favelas, we realize that, at least in the case of Rocinha, it is the last case described by Olivier de Sardan that best designates the first enterprise of intermediarity capture. It is necessary, however, to modify the typology a little, since this, in the case that corresponds best to Marcelo, evokes a will to leave the local political arena with the aim of an external ascension. However, Marcelo has never entered a local structure and is not a local in any way. When we listen to his life story and observe his current situation, more than 20 years after the first tours, it is clear that, for him, the interest in this activity concerned the success and external recognition of his company, which provided him with the essential part of his economic capital as well as of his social prestige. Moreover, it also appears that his intermediarity leans more towards the tourist than the local population, a tendency that can be clearly explained if we consider that his main objective is that of a social ascension outside the favela. Marcelo is only really known in the favela by a few people who regularly encounter the tours. Finally, the case in which we can classify Marcelo as a cultural broker also largely explains the very structuring of his tours, and in particular the almost total absence of interaction between tourists and locals during the tours. Another example, that of Zezinho, in a completely different case, illustrates how the content of the tour depends on personal trajectories.

3.1.2. *Zezinho and Tony, the favela from the inside*

Since the beginning of the 2010s, the number of independent guides offering visits of the favela has greatly increased, and some individuals have been able to create tour operators that now manage to compete with the pioneers. Zezinho, founder of Favela Adventures and based in Rocinha, is one of them. While his case is particularly revealing of this propagation of a valorization of the local in the development of this tourist activity, it is precisely because he himself is a character straddling two worlds, and whose efforts to make the tour as local as possible are particularly interesting to illustrate the appearance of a new type of actor.

Zezinho is actually American-Brazilian. His parents (American mother and Brazilian father) met in the favela, while his mother served in the Peace Corps, which was very active in the favelas of Rio de Janeiro in the 1970s[3]. He grew up in California and did not return to Brazil until 2007, where he first worked as a vendor at an artisan market for tourists, before starting his own business. Zezinho never recounted (at least in my presence) his experience as a vendor, preferring to emphasize his desire to return to Rocinha to "help his community". The fact that it is clear that he belongs to the favela is an important part of his identity. He is covered with tattoos representing the favela's houses, and the words "favela" and "Rocinha" are written, like graffiti, on his arms and thighs. He proudly wears an array of T-shirts representing the favela. For some years now, he has had plastic bracelets made abroad with the inscriptions "I love Rocinha" or "Eu amo Rocinha" to distribute to the inhabitants of the favela and the clients of his tours.

Zezinho is more like the first case mentioned by Jean-Pierre Olivier de Sardan. At the head of one of the most active tour companies in Rocinha, he successfully developed his company during the 2010s. Everything in the structure of his tour

---

3 Licia Valladares (2006) returned to this period of Peace Corps activity in the favelas of Rio de Janeiro.

suggests that the acquisition of an intermediary position constitutes for him a strategy of local social ascension, and even more simply of local positioning, even though he has lived a large part of his life outside of the favela. Zezinho's tours clearly reveal the double play of this type of intermediarity, since it is largely through excursions that Zezinho simultaneously deploys his role as intermediary with tourists and his strategies of local implantation. This double aspect of Zezinho's activity explains the centrality of the empowerment logics he favors for his tourist enterprise. Unlike Marcelo, Zezinho places himself in a position where he is accountable to the favela's inhabitants. Realistic, like most of the guides, about the relative importance of the financial dimension of the activity, it is through the affirmation of a positive cultural identity of the favela that Zezinho intends to contribute to local life. His investment in recognizing the favela's culture is strong. It is first symbolized by his appearance and his body covered with tattoos in honor of the favela. Then, in parallel to his activity as a guide, he has become the main promoter of a love for the favela, materialized by his bracelets "Eu amo Rocinha", which passes above all by the affirmation of self-esteem, and not only by the development of an external recognition. Because of this double play relative to his objective, Zezinho's position, which I have had the opportunity to observe for several years, is much more evolving than that of guides with a more external posture and who are not engaged in local social ascension strategies. By taking advantage of the apogee of the favela tours, Zezinho has become a local personality whose network (notably the one that captures part of the development resources that Olivier de Sardan speaks of directly) has considerably expanded.

Zezinho is very critical of other tour operators' activity and has built his tour in opposition to the initial model, especially to Favela Tour. For him, the almost exclusive use of motorized transport to visit the favelas is inadmissible, and these tours are, according to him, akin to a "human safari", an image that is widely spread in the ethical polemics on the favela tours. He does not own any buses and uses public transport to pick up the tourists in the seaside. The tour then takes place entirely on foot, through the alleys of Rocinha, and includes a lunch in a favela restaurant. Zezinho's clientele is much younger than Marcelo's. They are looking for a more personal experience, and above all trying to avoid participating in a tour that they consider reprehensible or that could lead to their association with a type of tourist from whom they try to distinguish themselves.

For Zezinho, his action is part of a "sustainable tourism", which he confirms as "100% local". In fact, he often employs foreigners as guides, but requires them to live in the favela and encourages them to show their love for the community. Tony, one of the foreign guides who began his guiding work with Zezinho, is a good example. Originally from Cornwall, he arrived in Rio de Janeiro in 2014 to follow his girlfriend on a university exchange. He met Zezinho through a friend, and became close through a shared passion for DJing. Indeed, Zezinho founded a DJ school in Rocinha that he finances, according to his statements, with 40% of the income generated by Favela Adventures. It is often around this activity that he meets young Americans or Europeans (rarely older than 35), eager either to find an occupation during their stay in Brazil or to go there for humanitarian work. Having broken up with his girlfriend, but wanting to stay in Rio de Janeiro, Tony naturally saw Zezinho's business as a way to fulfil his desires. So, with the feeling of having had a positive experience, he moved to Rocinha and started working as an English teacher for a tutoring association advised by Zezinho. Tony quickly changed directions and turned towards becoming one of Zezinho's favela guides, following his proposition, rather than seizing the opportunity to participate directly in the DJ school founded by Zezinho, which operates intermittently. For almost two years, Tony has been one of Favela Adventures' most active guides and has acquired solid skills, much needed to counterbalance the frequent surprise of tourists who, having read the advert unambiguously talking about the guides as "100% residents", are met with a young blond man with blue eyes and a southern English accent. The origin and the accent, moreover, are not the only factors of doubt as to the legitimacy of this guide. Indeed, although he has made significant progress over time, Tony started his favela career with a very low level of Portuguese, and tourists often notice that locals, when interacting with him during a tour, perceive him as a foreigner and often find it difficult to communicate with him. Most of the time, Tony manages to get out of the discomfort that this type of situation can create by his good humor and his sense of self-mockery, which allow him to establish warm relationships with the inhabitants he meets and who see him, day after day, leading the tourists through the alleys of the favela.

While they are indeed required to live in the favela in order to become guides for Favela Adventures, the guides chosen by Zezinho do not have to be indigenous. On the contrary, it seems that the list of guides employed over the years by Zezinho suggests, with a few exceptions, a preference for a certain profile: male, aged between 25 and 35, Western and ready to settle in the favela for a long period of time. Also, the recruitment method chosen by Zezinho does not seem to be far from that of Marcelo and, in fact, of the vast majority of tour operators in the sector. Indeed, it is almost systematically the case that individuals who manage to take on the role of a guide have this profile of an international background, which is widespread in all forms of international tourism. This is true, with rare exceptions, for Brazilian guides who, for the most part, also have a path that has taken them to Europe or the United States for an extended period of time, from where they have returned not only with a decisive knowledge of Western mentalities, but also with solid linguistic skills. Also, although it is emphasized by some, the guides' membership in the local society is not a fundamental criterion. Even Zezinho, who distinguishes himself from the competition by his insistence on the local dimension of his tour, does not seek out favela residents at all costs and is easily satisfied with turning guides from outside the favela into residents.

Tony's case also corresponds to the type of strategy described in the first case of Jean-Pierre Olivier de Sardan's typology, although unlike Zezinho's, in whom the degree of premeditation is much greater, his strategic posture has evolved over time. Indeed, the beginning of his guiding activity is not far from being a complete coincidence, and the strategy that supports it is not initially finished. After two years with Zezinho, Tony chose to create his own agency. More than for the possibility of increasing his income, it was for greater control of the redistribution of the tours' profits that Tony founded his own tour operator. Indeed, the tours he takes care of now allow him to finance the creation of his English school in the favela. Today, he even employs other guides from Rocinha in his business. What is special about Tony's case is the length of his involvement in the favela, which has thus extended beyond Zezinho's grip. Tony seems to embrace his way of life in the favela, which offers him a certain independence. Indeed, Tony is illegally living in Brazil, where he has overstayed his tourist visa for several years now[4]. It was the prospect of this lasting settlement in the favela that was undoubtedly one of the main motives for Zezinho hiring Tony, who clearly wanted to be able to tell the truth when he cleverly played on the idea of his guides' indigenous character by using the term "residents". It is only that Tony's commitment has become more enduring over time, and it is clear that his choice of permanent residence in Rocinha has had a significant impact on the way he approaches his intermediality. However, the simple fact of residing in the favela, a quality required by Zezinho, inevitably leads the guides into a strategy aimed at their local integration.

Unlike their external competitors, the local tourist agencies are not, in Rocinha, structures where guides are employed exclusively and permanently. In fact, the companies created by local people only employ guides as needed, and in no way prohibit them from working for others or on their own. Here, it is the difference in the status of guides that comes into play. It is very rare for local guides to have an official guide card. The work of guides is therefore mainly informal at the local level.

The local guides must possess two qualities, both of which are important: an indigenous capital and a command of English. What is emphasized by local guides is their belonging to the community, in that it allows them to be situated in opposition to non-local agencies. However, this "indigenous capital" can in fact be fabricated, as Favela Adventures does, so that individuals from outside the favela but who are highly cosmopolitan can be assimilated to locals. In this, as with non-local guides, local guides are chosen primarily for their ability to interact with tourists in multiple languages and to demonstrate a keen understanding of their expectations. Local agencies are often small structures, in most cases composed only by their founder who is also the main guide. Most of the time, only in the case of excess demand or

4 He has since left Brazil.

unavailability, other guides are employed. The range of languages offered to tourists is smaller, and it is rare that tours are organized in languages other than English. During the Olympics, some local companies employed other guides for their language skills. For example, during this period, there were tours in Dutch or French provided by companies claiming to be local.

3.1.3. *Obi, Erik and Paolo, the indigenous guides*

The second case evoked by Jean-Pierre Olivier de Sardan, that is, the case where the individuals, before acquiring a position as an intermediary, are locally dominated or marginalized characters, could also correspond to Zezinho if we did not consider him as a character coming from outside. In the case of the tour guides, it does not seem possible to identify characters who would be in a locally dominated position. However, it is obvious that it is through the activity of a guide that most of them find their first opportunities to insert themselves in the local public sphere, and then ascend towards acquiring prestige. Also, if they are not voluntarily marginalized, they find themselves in a marginal position vis-à-vis the main political structures of the favela. In reality, the possibility of becoming an intermediary through becoming a guide offers the opportunity to make use of a capital that is not valued in the classic structures. Here, the tourist opening of the favela allows local individuals, such as Obi and Erik, to use their linguistic skills and cosmopolitan capital for the purpose of local ascension. As with the guides in the previous case (Tony and Zezinho), their tours are structured not only for the tourist, which makes them legitimate mediators, but also for the locals, since it is through the tours that they can let the locals see how they do their guiding and sometimes build a network. Thus, local guides have a double undertaking of legitimization. While it is sometimes the norm in other favelas (notably in Santa Marta where community tourism is strong, as well as in Morro da Providência, which seems to have only local guides), the indigenous nature of the guides is, however, rare in Rocinha. Nevertheless, it has become increasingly frequent, especially with the arrival of a new generation of guides, as shown by the examples of Obi and Erik, both born in Rocinha.

Obi is a young man, aged 25. His brother, who has since immigrated to Australia, founded one of the few youth hostels in Rocinha in their family home. Since his older brother left, Obi has been running the Rocinha Guesthouse with the help of his mother. Part of Obi's small house has been transformed into a hostel, with Obi sometimes sleeping on the couch in the living room when guests occupy the dormitory that his old room was turned into. His work as a hostel manager has allowed him to acquire very good English, which he says he "learned by talking to tourists in Rocinha". It was because he saw his business as proof of his inventiveness (and, by extension, proof of the innovativeness of the people of

Rocinha) that Zezinho was attracted to Obi's profile and offered him a job as a guide. Although Favela Adventures provides him with the bulk of the tours, Obi has never worked exclusively for Zezinho, and has long offered tours to clients of his hostel, while sometimes working for other agencies. In 2017, he founded his own tour agency, Favela Brothers Tour, which does not, however, provide enough business, forcing him to sometimes work for Zezinho and recently for Tony. As he trained as a guide with Zezinho, his approach to the tour is not very different: it is based on the demonstration of the adaptability of Rocinha's residents and on the staging of an alternative living environment, all of which is enhanced by information similar in every way to that given by Zezinho.

Erik, who is the same age as Obi, is also from Rocinha but has a different life story. Among the Rocinha guides who emphasize their locality, he is in fact one of the only ones who can claim to be a tourism professional, since he studied tourism at Senac[5] in Rio de Janeiro. However, he has been influenced by several tourism personalities in Rocinha. When he was a young teenager, he took English classes at a school funded and frequently visited by Rejane Reis, founder of Exotic Tours, one of Favela Tour's historical competitors. For Rejane, Erik's example is testament to the success of her community work. However, it was with Zezinho that Erik really started his guiding business. This did not stop Erik from founding his own tourism agency, Rocinha by Rocinha. In recent years, Erik has become one of the most sought-after guides in the press and often speaks on behalf of local guides. Obi and Erik represent a new generation of guides, born in the favela, who have seen in the tourist development of Rocinha a career opportunity. Although they arrived at their profession via different paths, they both were influenced, at some point, by some major figures of tourism in Rocinha, and in particular of Zezinho, who saw them as considerable assets, notably from the guarantee provided by their indigenous capital. It is likely that if Obi and Erik were to apply to Favela Tour, their profiles would not be selected by Marcelo, who not only does not attach particular importance to the guides' locality, but is even wary of it, believing that a discourse too centered on Rocinha would be detrimental to tourist expectations.

Before we go on to talk about another type of guide, it is necessary to introduce Paulo Cesar "Amendoim". Often presented as the oldest guide in the favela, Amendoim is indeed a pioneer. He too has taken advantage of the favela tour boom by intensifying his activity, helped by his status as a pioneer and a native. Paulo is very well known in the favela, where he was president of the inhabitants' association, and no other guide can claim to know as many people as he does in Rocinha. However, Paulo has a similar background to most guides. In fact, Paulo, in addition to his activity at the head of a residents' association, has experienced an international career that began with his work as a shoeshine boy in a large hotel in the city, which

5 National Commercial Learning Service.

provided him with his first clients "long before the favela tours". Then, it was an athletic career that led him to travel around the world during a part of his young adult life. In this, Paulo's case, even as a representative of an intermediarity acquired through the extension of a local status, does not therefore allow us to dismiss the idea of an international life trajectory as a fundamental common point that unites the different guides, whether or not they are local, or whether they have acquired cosmopolitan skills virtually, through travel or through meeting with tourists in their home towns. Amendoim is nevertheless one of the two or three individuals who could be classified in the third category of cultural brokers described by Olivier de Sardan, rarer in the case of favela tours. Indeed, Amendoim, sometimes considered by local guides as one of the oldest guides of the favela, represents the case of acquiring intermediarity from a position already central in the community. Unlike Zezinho, he did not have to work simultaneously, over the course of interactions during the tour, on his position as a guide and his place in the community. Indeed, Amendoim can count on a local celebrity posture that stems from several events. First, he was president of one of the main *associacão de moradores*[6] for eight years. Then, in the early 2000s, his participation in *No Limite*, the Brazilian adaptation of the American reality show *Survivor*, made him famous for a while in the country and definitely propelled him to local celebrity status in Rocinha. When observing his tours, it is clear that the presence of tourists around him is a way to reinforce his social prestige. Sometimes, much to the tourists' dismay, his tours are filled with personal interactions with the favela's inhabitants. However, his journey remains relatively rare and does not allow us to generalize the process that leads from a posture of local prestige to the acquisition of an intermediary status.

3.1.4. *Alex and the independent guides*

Tours can also be organized by individuals from outside the favela, who offer favela tours completely independently. As in the case of agencies, the distinction between local and non-local is also relevant to differentiate independent guides. While some have an entirely independent background, it is common for these guides to have started their activity in the favelas by working for local agencies, particularly in the case of indigenous guides, or outsiders. There is, on the other hand, a certain continuity in the ways of visiting, with foreign guides sometimes continuing to use motorized vehicles (cars for small groups). However, there are some exceptions, as Alex's case shows. Alex, a former French employee of Favela Tour, shows visitors around several of the city's favelas, giving preference to walking tours. For him, leaving Favela Tour was, in some ways, liberating:

---

6 Basic associations in the favelas of Rio. Due to the considerable size of the Rocinha community, there are several associations.

> Alex: I'd much rather do the tours like I do than sit on the bus all day. You've seen my tour in Cantagalo, right? It's great, you see a lot of things, you walk around. Honestly, I go everywhere without hesitation.
>
> Me: And Rocinha, do you ever go back there?
>
> Alex: Sometimes, when people really ask. But I don't really like it, I don't know it that well and it's too big (Interview with Alex, August 2015).

Visiting the favelas remains one of his main activities and he has a solid clientele, to which he owes two things. First, Alex is registered on the official list of French-speaking guides of the French Consulate. This mention assures him a French clientele, which constitutes the importance of the audience of his agency Rio Autêntico. Alongside his work as a guide, Alex has built, with his French friends, a large house in the small favela of Pereira da Silva which, since the 2014 World Cup, has welcomed tourists. This activity also allows him to meet potential clients for his tours. However, Alex maintains a fairly clear distinction between the favela he lives in and those he tours:

> Alex: We wanted to move here because it's a small community. We have an incredible view of the bay, you know? It's really cool, it's a dream.
>
> Me: But don't you show the favela?
>
> Alex: Here? No! Not unless they really ask for it. But first of all, there's not much to see, and secondly I prefer to avoid it. We're good with the locals here and if we start walking around with tourists all day, it'll get on their nerves, especially the guys at the bottom [drug dealers are stationed at the bottom of the favela as we speak].

3.2. The privilege of mediation

According to the guide profiles, the stakes of obtaining mediation are different and correspond to various hopes for its conversion into profit. The interest in having drawn out, through these examples, a typological distinction, lies in the desire to highlight the importance of determining the sphere in which the strategies of each are deployed, in order to understand the ways of structuring the excursions. Such an approach is essential in order to understand, for example, how private interactions between guides and local people, which multiply during the tours, to the dismay of some tourists, do not have the (sole) objective of "playing local" in front of the tourists, but rather respond to issues of integration and local legitimization, the importance of which we may be tempted to underestimate.

Even before seeing the way it is thought and worked by the tourists, the distinction between local guides and non-local guides must be analyzed within the framework of this typology of the various profiles and ambitions of the actors who, through tourism, acquire a position of intermediarity.

3.2.1. *New political intermediaries?*

Often perceived as a bubble, an activity totally disconnected from the rest of social life, tourism is in fact a phenomenon that goes far beyond the framework that is usually set for it. This is the case, in particular, of the different symbolic points that underline favela tourism and structure the discourse of the guides. I will show in the coming sections (sections 4.1 and 9.1) the importance of tourist narratives of their experiences. These also allow the discourses received in the favelas to spread outside, thus feeding the Western imaginary which, with respect to these discourses, is thus both a source and a product. But there is another phenomenon worthy of comment here, directly concerning the guides, and showing how they can sometimes become mediators between the favela and the outside world in limits wider than those of tourist activity strictly speaking. Indeed, during the organization of mega-events in the city, many journalists have turned to tourist guides, almost exclusively local guides (although there are some exceptions), to gather information about the favelas.

I noticed this fact in 2016, a few weeks before the Olympic Games, when a Carioca activist, Theresa Williamson, organized a media tour in two favelas (Vila Autódromo and Cidade de Deus) and a quilombo (Camorim) for journalists who came to cover the event. In two destinations (Camorim and Cidade de Deus), the interlocutors chosen by Theresa were tour guides. The nature of the relationship between visitors and guides, as well as the discourses, did not appear to be different from what I was used to observing in a tourist context. Thus, I thought that the tourist model for presenting the favela had influenced the type of visit chosen for this media tour. Indeed, the guides intervening here quite naturally reproduced a type of discourse elaborated in their work as a professional guide, and thus represented the human continuity in the diffusion of the discourse from a tourist framework towards a wider media framework. They were the identifiable actors influencing tourism on the formation of media treatment.

In addition to the importance of their presence in a media tour on the formation of media discourse, the use of guides also reveals the emergence of a new important political actor in the favelas. In recent years, the guides have become the favored interlocutors of foreign journalists[7]. It is not uncommon to find journalists on favela tours, although most often claim to follow an individualized tour led by the owner of the tour operator. Although in this type of situation the guide has a certain influence on the information transmitted to the journalists, the latter seems to be most interested in the tour for the opportunity it represents in entering the place, photographing it in complete safety, meeting the inhabitants, etc. In addition, although their purposes are different, tourists and journalists have a similar relationship with the tour, particularly insofar as the guide is, at first sight, more of a go-between than an actual informant. However, in both cases, the guide has a much greater influence and power of control than the recipients of his discourse assume. And the role he has gradually taken on for the international media is not insignificant.

This place is not limited to the participation of journalists in tours. In fact, guides have now become key interlocutors and, in particular, during periods when global events are being hosted, they are one of the most heard voices in international media. This is a fact that I have observed episodically for years, and that really became systematized during the period of the Olympic Games, when most of the guides and tour operators owners were solicited daily for interviews, visits for journalists, etc., so that it was more difficult for me than ever to obtain their time to pursue my investigation. In Rocinha, for example, no other individual was interviewed and photographed as much as Zezinho, who could be seen almost daily on the bridge leading to the entrance of the favela, accompanied by a film crew or a journalist handing him a microphone. It is also clear that the tourist guides have taken place in the media space that no other intermediary or representative of the favela had managed to take before. In this way, they replace a number of traditional political actors, such as the *associação de moradores*, who are then removed from the channels of direct communication and made dependent on the intermediation of the guides. Therefore, it is evident that this is accompanied by a modification of the discourse. In terms of media content, the discourse is thus influenced by mediation. The most important impact of this displacement of the guides into global media interlocutors lies on the scale where the favela is discussed, which is thus staged primarily for the otherness it embodies. The primacy given to the guides for the presentation of the favela in terms that correspond to Western categories, and thus to the detriment of the inhabitant' associations, implies that the favela is represented as a whole, as a relevant general category. This is undoubtedly the main difference between the guides' discourse and the one that the leaders of the inhabitants'

7 This is not the case in the Brazilian media, which did not penetrate more the favelas during the Olympic Games, and on the contrary did not address, as we will understand quickly why, the guides.

associations could give if the opportunity arises. Indeed, the latter are involved daily in projects and issues that concern their neighborhood; they register and relay the complaints and demands of the inhabitants. Also, the discourse that these actors could present is certainly more focused on the problems and specificities of their favela, and not on a vague category that would include all favelas.

But how did the guides come to occupy this position as intermediaries between their communities and international journalists? Guides have probably become the interlocutors of choice for international journalists in the same way that they became guides in the first place, that is, by the way they respond to the foreign tourists' expectations. They are individuals with great intercultural skills, primarily linguistic. They know how to balance information in order to reintroduce elements of a preconstructed image, while at the same time surprising their interlocutors by maintaining the effect of discovery. This is mainly linked to the international nature of their background, and many have progressively acquired an important media dimension because of their origins. Indeed, it is quite common for favela tour guides to be nationals of other countries, particularly in Europe or the United States, who have settled in a Carioca favela. In the same way that their double nationality was originally able to provide them with a clientele for the tourist activity, they frequently become the favored interlocutors of the media in their country: consider Alex for France, or Tony for England.

On the other hand, these guides are specialists in the dissemination of a somewhat paradoxical discourse, which consolidates the specific interest in the favela while allowing them to maintain the impression of the absolute necessity of their position as intermediaries. On the one hand, the discourse of the guides allows them to support the valorization of the favela, and thus a certain enchantment of otherness. On the other hand, it makes it possible to highlight the distance to where the favela is located, to give not only to the guides the decisive argument to ensure the perpetuation of their roles, but also to ensure their addressees of the possibility of drawing a certain prestige from their visit, to give to their discourse an argument of empirical authority (particularly important as much for the journalists, but, as we will see, for the tourists too), while risking the feeling of an authentic and exotic experience. This feeling is present in the media discourse emerging from these situations.

3.2.2. *Away from mediation*

The mediation between tourists and favelas is thus carried out by a group of individuals who have managed to gain intermediary positions by way of enhancing their profile. As the examination of the guides' biographies shows, access to mediation depends more on intercultural communication skills than on a possible indigenous status, which in any case is not enough. This struggle for access to mediation is all the more important because tourism in the favelas is structured around the figure of the guide, whose presence is an absolute necessity. Most of the arguments I have put forward to comment on the means of access to the activity of being a guide in the favelas have focused on the relationship with tourists, and this is why intercultural skills are the most important. The guides enjoy the essential means of presenting the favela, but they also condition the opening and limitation of a space where the inhabitants who hope to benefit from the presence of tourists can insert themselves.

During the excursions, the inhabitants are most often confined to being onlookers. However, they can join the tour in many ways, starting with occasional interactions with the group. Most of these interactions take place between locals and guides, when they greet each other or, in the case of local guides, exchange information about daily life affairs. When a local and a tourist engage in an interaction, it is most often on the fringes of these discussions with the guide and on a local initiative. These interactions do not escape the guide's mediation, who must translate these conversations. This linguistic discrepancy is the only reason why interactions between tourists and locals are rare. Most of the time, the inhabitants show a general indifference towards the tourists.

However, some residents take advantage of the favela tours. In addition, all the inhabitants hope that this possibility will be open to those who wish to seize it: one of the main benefits expected from tourism. In practice, however, this possibility is not open to all, and remains subject to the discretion of the guides who control the incursions of local people in the tours. This concerns both the inhabitants/traders, whose activity is not by vocation touristic (the restaurant owners or the associations), and those who specifically undertake an activity intended for the tourists (performers, craftsmen/traders). As said before, the guides have two audiences in front of them. The first is that of the tourists, their clients, for whom they ensure proper conduct and quality of the tour. In this perspective, they are led to control the incursion of others and to operate a selection at the same time quantitative and qualitative. They first make sure to control the number of solicitations and often limit access to the group to one bar and one craftsman/tradesman. The second audience is that of the inhabitants, who observe the tours and sometimes solicit the guides so that the excursions allow them to sell goods or performances. Allowing them to carry out their activity is important for the guides and, in a different way according to whether or not they are inhabitants themselves, for the durability of their tour and their mediation. The authorization to access the group is most often given on the basis of a tacit partnership by which the guide promises to make his tour pass by a craftsman's shop, to finish the tour with lunch in a restaurant or to visit (and contribute) to a specific association. These partnerships limit the number of solicitations and exclude residents who have not made such agreements with the guides. But they also have a qualitative dimension, since the guides choose, on the one hand, according to their affinities and, on the other hand, according to principles that they believe are shared by the tourists, but about which they often have different visions. This is what the following case shows. In Rocinha, the local tours very often include a moment at the bottom of the buildings constructed under the Programa de Aceleração do Crescimento (PAC)[8], to discuss the various urbanization projects that have affected the favela. A ten-year-old girl lives in one of these buildings. When groups arrive in the area, she rushes to meet them and offers to sell them magnets representing the favela with a few words of English. Some of the guides have positive opinions about this request:

> Solicitations are the name of the game. For example, that girl we saw earlier who sells magnets is great, she's always smiling. In general, people buy and I am happy that she can sell these magnets that she makes herself (Local guide, Rocinha).

For others, on the other hand, the age of the vendor is problematic. Some even avoid passing by, as well as other performances also made by children:

> I'm very careful around the kids, especially because I know it can shock tourists – child labor, all that. For example, this girl who sells magnets. Well, she's very nice, but her parents push her out when the tourists are there. They know that for some tourists, it works better. I prefer to avoid that (Local guide, Rocinha).

As for the associations to which the tours contribute and which the guides show tourists, the selection is of the same type. Sponsoring a social association, most often a school support association, is important for all the guides, since it is essential to be able to present the social impact of tourism in the favelas. It is also important to be able to visit them, even though it means interrupting classes, because this visit is the only proof of the tours' contribution to the association. The guides therefore form partnerships with associations. Depending on the association's progress, these partnerships can evolve and even break down:

> You've seen, since last year, we changed associations. The one we used to visit was old and not doing well. And every time we came, we

8 The PAC program is a favela urbanization project launched in 2007. In Rocinha, the main results were the construction of a footbridge designed by Oscar Niemeyer and the renovation of rua 4, whose walls are now colored. Residents have often expressed their disappointment with the PAC, and have questioned the priorities of the federal government responsible for the program. See: https://rioonwatch.org.br/?p=2191.

> saw no one, not one child in class. It gave a bad image, so we decided to stop and change (External guide, Rocinha).

Whether they are shopkeepers, craftsmen, restaurant owners or performers, the favela inhabitants can only access tourists through partnerships with the guides. No other form of decision except the sometimes arbitrary one made by the guide is taken into account in the choice of activities open to the inhabitants to benefit from the spin-offs of tourism in the favelas. This participation therefore depends on the ways in which the guide has acquired the benefit of tourism mediation and what he intends to do with it, either to improve his position in the social space of the favela, or outside of it. The avoidance that characterizes the interactions between tourists and inhabitants (and which I will discuss in Chapter 9) thus stems in part from this phenomenon. In these conditions, it is indeed difficult for the inhabitants to influence practices and to find an active position in them. This applies to the creation of commercial opportunities as well as to the expression of reservations about the content of the tours, and it also reduces the space in which the inhabitants can show their hospitality. To conclude this chapter, however, it is worth presenting the exception of a resident's work to become an active hospitality professional.

It was with Erik that, for the first time, I discovered the Portal Joanas. The existence of the place had already been pointed out to me through discussions with other guides, but I had never had the time or the opportunity to go and see for myself what it was really like. It was therefore with great satisfaction that I learned, at the beginning of Erik's tour (Rocinha by Rocinha) that we would stop with the group (two South Africans, two French and three Italians) in this place that he presented simply as "favela house". The visit took place towards the end of the excursion; it was almost noon. A few meters from the estrada da Gávea, we stopped in front of a house on rua Um, located behind a gate. Erik rang the intercom but there was no answer. He rang again. After a few seconds, a young woman came out and presented herself at the top of the small stairs that lead from the gate to the door of the house:

> Young woman: Oi.
>
> Erik: *Bom dia*, is your father there?
>
> Young woman: Yes, he's getting ready. Wait a moment.

Erik then turned to the group and began to explain what was going on: "You are going to meet César. César is letting us visit his house. I warn you, he's a character!" Barely a few seconds passed, the gate door opened and we followed Erik up the stairs. We arrived in front of a wooden door that opened immediately. It was César who opened it for us. César was about 60 years old and his clothing indeed suggested an original personality. He was wearing a Brazilian national soccer jersey, a jester's hat with bells and fake plastic glasses, both in the colors of the Brazilian flag. Right away, César showed amazing energy and enthusiasm. He introduced himself to the tourists one by one, greeting them with a frank *abraço* for the men and a kiss on the hand for the women, while repeating dozens of times, "Bem-vindo, welcome to Rocinha!" Not speaking English, it was through Erik that he inquired about the origin of the tourists, not failing to show a playful fascination: "Ah França! África do Sul!" Throughout the visit to César's house, which he named Portal Joanas, Erik turned into a simple translator, enjoying a moment's rest for a while.

César offered everyone a coffee or a lemonade, which his daughter prepared behind the bar of the American kitchen that divided the large room where the tourists were welcomed. César, meanwhile, was delighted to show the tourists his house and the few gadgets that were in it. With obvious amusement, he showed that his glasses started blinking in yellow and green thanks to the operation of a small button located on the arms. He was also very proud to show a collection of foreign banknotes, which decorated the bar under a glass plate, with some columns around it.

He repeated, while drinking coffee, some obvious facts already well known to tourists, and which are real tourist slogans of Rocinha: "You are in the biggest favela of Brazil," "here, life is beautiful, it is a happy community", etc. In the background, the music consisted of a mix of great Brazilian hits: Jorge Ben Jor, Tim Maia, Gilberto Gil, Chico Buarque, etc. When "Garota de Ipanema", one of the most famous Brazilian bossa nova standards, played, he turned up the volume and started to dance, inviting the women one by one to join him so he could teach them some samba steps, in front of an amused look of the rest of the group. Once the dance was over and the coffee was drunk, César invited everyone to follow him into the hallway and climb the stairs to the roof. César's house had a breathtaking view of the favela, and it was with obvious pride that he shared it with the tourists. As everyone took out their cameras, César launched into a description of the Pedra Bonita mountain, which lies opposite the favela, to the left of the sea. The impeccably sculpted relief of Pedra Bonita has given rise to many more or less fanciful theories that César evoked, without omitting of course the popular hypotheses relative to an extraterrestrial intervention.

While the tourists were still busy taking pictures, César asked Erik to repeat to him the origin of the group members; he then disappeared for a few seconds. Returning from the house, César brought some pieces of cloth that he put next to the door providing access to the terrace; then, he attracted the attention of the tourists and invited them to take a group photo. It was his daughter who took the picture. Before entering the frame, César unhooked a Brazilian flag from the wall and wrapped himself in it, as he stood in the middle of the tourists, in front of the view of the favela, then raised his arms to the sky while unfolding the flag. For the second photo, César asked everyone to raise their arms or jump at his signal, while he threw the flag in the air. Then, he insisted on taking a picture with each group according to their origins. He then asked his daughter to go and get him the cloths he had just pulled up. It was a French flag and an Italian flag. After apologizing to the South Africans for not having a flag of their country, he began to link the French and Brazilian flags for the sake of the Franco-Brazilian photo, to which I was logically invited. After having repeated the process with the Italians, we were invited to go back down to the house. There, César handed out a wooden board on which he had drawn the favela, most of the time next to another tourist symbol of Rio de Janeiro. It was a work of pyrography. To each of the tourists, he asked them to momentarily return the work so that he could put their first name on it, thus demonstrating his talent as a pyrographer. After the handing-over of the gifts, and when the group was ready to take to the road again, Erik invited the tourists to leave, if they wished, a small donation to César. That day, as on all of the days I subsequently accompanied a group to César's, almost the entire group gave a minimum of 10 reais. Delighted by their visit to the interior of a favela house and their meeting with this atypical and warm character, the group left the house under the greetings and friendly embraces from César.

César's initiative is unique on several levels, and few inhabitants have succeeded in taking advantage of the tourist presence in Rocinha. If he enjoys today the tourist development of his district, it is because he brings an estimated value to the excursions. He has contributed in various ways. First of all, it has an important element, as much for the local guides as for the independent ones who work regularly in Rocinha: a terrace with a view. In this respect, the fact that César opens up a house is not unique. Indeed, many inhabitants of the upper part of Rocinha have entered into partnerships with the guides, thanks to which they secure a regular income in exchange for opening access to their roof, and thus to a viewpoint overlooking the favela. Favela Tour, for example, has been using the same terrace for more than 10 years and considers it to be an indispensable step in the excursion. In their case, the terrace is accessible through a garage. It is different for César, and this is precisely the reason for his success. Indeed, César combines access to his terrace with a passage through his home. It is a great thing for the guide to be able to offer tourists the opportunity to enter a favela house. This is especially true for independent guides who come from outside and who, during their tours, do not have many opportunities to create links with the inhabitants. In any case, it is relatively rare, in Rocinha as in Rio de Janeiro in general, to be invited by someone into their home, and appointments often take place outside, in bars or public spaces. For example, for Céline, a French freelance guide, the visit to Portal Joanas has become a systematic step:

> Yes, that's for sure. For us, it's great to go to César's, you can see a house from the inside, the tourists are very happy. Plus, he's an exceptional guy (Cécile, independent guide, interview June 2017).

But it is not only foreign guides who see César as an asset for their tours, and local guides also drop in regularly for several reasons. For Erik, with whom I met César for the first time, the existence of Portal Joanas is almost as decisive as it is for the outside guides. Indeed, Erik still lives with his mother and siblings, and does not plan to give daily tours of his home. In general, local guides are quite resistant to the idea of opening the doors of their own homes. If they sometimes do it and regularly did it at the beginning of their tour, they seem to get tired of welcoming tourists every day to their homes. On the other hand, César, as you will have noted in the description above, is not satisfied to only open his doors: he also shows off his personality. It is he, in fact, who creates interest in Portal Joanas, through the way in which he offers warm hospitality on which the guides can rely to transmit to the tourists the idea of a friendly and receptive population to tourism. By talking to César and observing the way he greets tourists on a daily basis, they can understand quite well the perception he has of his role. Far from being as strategic as guides and tourism professionals can sometimes be, César has two expectations of himself. The first, and most important, is the warmth of his hospitality. César spares no effort when welcoming tourists to leave the impression of genuine joy and exceptional dynamism. Having never had the opportunity to witness a different attitude, it does not seem to me to be an exaggeration to say that his behavior and his good humor are genuine, and that his new occupation, when he thinks of his former life in the factory, brings him real satisfaction. The second element that César feels obliged to bring, and through which he distances himself from the discourse of the guides, is an exacerbated Brazilianity. His clothes, accessories, the decoration of his house, the music he plays when the tourists are there: everything is designed as a reference to a patrimonial Brazil. The reason why I say he diverges in this respect from the guides is because the latter generally insist on the favela as a specificity and rarely refer to a purely Brazilian imaginary. In this, César designs a first-degree encounter. Imagining that the tourists come to meet Brazil and Brazilians, he dresses himself in the colors of the flag and exaggerates his mastery of Brazilian cultural codes. It is quite intriguing to note the success of this staging. Indeed, it seems that elsewhere, César's insistence on playing on the most obvious clichés could have constituted an obstacle to an impression of authenticity, or even to the feeling of hospitality, with the same tourists. But with César, the cliché becomes charming and is never a sign of a staged event. In fact, it is César's eccentricity that ensures the impression of great sincerity. The example of César offers a lesson on authenticity, insofar as it shows, in the heart of a tourist situation that is nonetheless alternative, that authenticity is much less attached to the symbols than to the one who stages them.

4

Distinguished Practices, Practices of Distinction

Favela tours and their development are the result of co-production. In fact, tourism in the favelas is not to be understood as a result of pressure from tourists, or as the fruit of its promotion by professionals. The two processes – supply and demand, if you will – are constituted together and are inseparable. This is how we can understand in part the importance given to the organization of favela tours for individuals with a cosmopolitan background; this gives them access to the knowledge of the symbolic elements that are essential to foreign tourists regarding favela tourism. In return, intermediaries influence the structuring of practices by creating a diversified offer, by distinguishing themselves from their competitors and by organizing a tour in accordance with the objectives they pursue in the social fabric in which they are embedded. Chapter 3 shows the influences on the diversification of practices in terms of their organization. It is now appropriate to consider the role of demand in this diversification, and thus to return to the tourists' side to observe the way in which they perceive the available offer and make a choice.

4.1. Criticism and distinction

To this end, I will discuss two things simultaneously. On the one hand, the strategies of distinction deployed in and around the favela tours by tourists. Here, I will present how visiting the favela can be useful for acquiring symbolic prestige. On the other hand, criticisms will be discussed, those made against the favela tours, and in particular the criticisms made by those who choose alternatives to visiting favelas. We will see that this double dimension, which at the same time makes favela tourism an opportunity for distinction and a risk of moral condemnation, hides in fact a complementary mechanism that constitutes one of the driving forces of the diversification of tourist practices in favelas.

4.1.1. *A valorized practice*

I recorded by chance, in Paris, a conversation between three individuals. Two of them, Marc and Jonathan, had more or less recently traveled to Brazil and had returned with different experiences of Rocinha. This conversation, provoked by my presence as an anthropologist studying tourists in the favelas of Rio de Janeiro, quickly turned into a debate about danger. This excerpt reveals the relationships maintained between the nature of each individual's experience (or for Paul, the third individual, lack of empirical experience) and the authority of the ideas and arguments presented:

> Marc: I went there at the beginning of the year, to Rocinha. I can say that at the beginning, I was not reassured. It is impressive. First of all, it is the biggest slum in South America. Well, slum, it's not really that, [looking at me] you know. [...] I have traveled everywhere, especially in South America, but here, I was a little apprehensive. And in fact, it was very quiet. There were guys with guns and everything. But in fact, no one cared at all. It's normal. Afterwards, what's interesting is that it's a whole organization, people manage to get water, electricity and everything. There is a lot of solidarity.
>
> Paul [turning to Jonathan]: And you, you went to Brazil too, right?
>
> Jonathan: Yes, two years ago. In 2009.
>
> Paul: And have you seen a favela?
>
> Jonathan: Yes, I already told Thomas. I went to Rocinha completely by chance. I was with a Brazilian friend and we got on the wrong bus. We arrived in front of the favela, we freaked out and we didn't go back, we just drank a Coke at the bottom, and then we left.
>
> Marc: Really?
>
> Jonathan: Yes. And on the bus there, I was pick pocketed by some kids from the favela. I didn't have any money left, and this made me very angry.

> Marc: Yes, you have to be careful.
>
> Jonathan: Plus, it was hot when I was there. The day I arrived, they had shot down a police chopper, though.
>
> Paul: Who?
>
> Jonathan: Well, in Rocinha, the gangs.
>
> Paul: Oh yeah, wow.
>
> Marc: Yeah, sure, it's crazy. But it doesn't represent reality, really. It doesn't actually take up that much space.
>
> [Undoubtedly challenged by Jonathan's doubtful face, Marc turns to me and resumes].
>
> Marc: Yes, most people work. They also party, they love their neighborhood, and everything. It's super welcoming, right?
>
> Me: Yes, yes, of course…
>
> Marc [interrupting me]: Yes, well, of course, you shouldn't idealize (Informal discussion in Paris, September 2012).

I had already noticed, even before I went to Rio for the first time, that those who visited the favela sometimes gained, if not prestige, at least a form of authority in the conversations, visible in this extract. The strength of the argument rests on a principle included directly in the favela tours, which is that of unveiling through the visit: going to the favelas being presented as the only way to glimpse at a reality hidden under the veil of prejudices that obstruct the knowledge of the favelas and therefore of Brazilian society. In this type of conversation, in which everyone has certainly had the opportunity to take part, the positions of each person clearly show the prestige conferred by the simple fact of "having seen with one's own eyes", and the superiority of the empirical over all forms of strictly indirect and encyclopedic knowledge. In this particular case, there are four distinct positions. Paul had no direct knowledge of the favela; he, as a spectator and questioner in the discussion, naturally did not express any ideas. He could have done so, however, if none of the other participants had had empirical experiences of the object of the debate themselves – in which case encyclopedic knowledge would be quite valid. Jonathan had already seen Rocinha, not only by mistake, but also with pronounced apprehension and a negative outcome. Again, if he had been the only individual with a direct experience of the favela, this would probably have given him sufficient authority, for example, to convince Paul of the superiority of his expertise. But the presence of Marc, who had a more recent and profound empirical experience than Jonathan, limited the impact of the latter's account, thus returning him to the position of a less competent tourist who had only seen on the surface what Marc had seen in depth. However, Jonathan's discourse is not to be discredited as Paul's could have been if he had expressed an opinion himself. The reason Jonathan's impressions retain some value is not because they appear to everyone to be particularly accurate or relevant, but because they are unquestionably empirical. Indeed, anything that is based on this type of knowledge is eminently incontestable and irrefutable, at least publicly. Unless we question the mental health and the good use of the senses of those who have seen, it is indeed impossible to contradict points of view based on an empirical experience. Moreover, the recourse to the regime of the sensitive, to impressions, to feelings, that is, to the display by one of the interlocutors of a subjectivity, is often valued. In the situation described, my presence also had an impact. Indeed, the nuance that Marc gave to Jonathan at the end of the extract, when he said that we "should not idealize", is certainly linked to this. Since I was not quick to agree with him, Marc protected himself from a rebuttal that may come from an interlocutor whose experience would likely be recognized by others as more profound. Although I made no attempt to position myself as an authority, the imagination of the anthropologist that exists among regular travelers gave me a superior hierarchical position here. The commented case thus offers a particularly interesting range of a hierarchy of travelers, unfolding from the immobile to the professional.

This first example, which could be generalized to all cultural tourism practices, shows how a visit to a favela has a value that is directly convertible into authority in a conversation. The tourist experience is transformed into knowledge, and this knowledge takes a dominant position in a debate of ideas. But this is not the only way to valorize the favela, and I will mention another, probably even more obvious than the first. Indeed, the mere fact of going or having gone to a favela is potentially a source of prestige, insofar as it allows us to assert ourselves as a particularly valorous type of traveler. Here again, small conversations randomly overheard, often far from Brazil, allowed me to realize this potential valorization linked to visiting a favela. It is perhaps in a conversation of this type that I situate one of the starting points of my work on favela tours. Not transcribed because it predates my decision to undertake an investigation on the subject, I remember that one of the protagonists had visited a favela, which he recounted with pride, and received the praise of the others who, for their part, tried to outbid each other by recounting the prowess of other trips to other countries, and above all to other places reputed to be dangerous and inadvisable to the class of "tourists", a class in which they obviously did not include themselves. It must be said that this type of people, young cultivated Parisians, represents one of the groups most attached to the valorization of the figure of the traveler and to the devaluation of the term "tourist", which they hate and consider to be an insult. Provided that the story is well told, visiting a favela is for them a means to reinforce this distinction and add to the symbolic capital that makes them travelers and not tourists. But the fact is not exclusive to one age group or social class, which presents only a particularly exacerbated version of it. The valorization of the traveler at the expense of the tourist is transversal, and runs through all leisure mobility practices, as it is almost never preferable to be assimilated to the latter.

As the favela is unanimously situated in a symbolic space that extends beyond that of tourism, it allows us to present the visit as being deeper, less stereotyped, less touristy in short. Often, visiting favelas is perceived and narrated as a side step during a stay in Rio de Janeiro marked by a more passive and trivial attitude, symbolized in particular by the moments spent at the beach and in the neighborhoods perceived as touristy:

> For me, it's crucial [to visit a favela]. You can't come to Rio and stay at the beach, hang out at Copacabana and ignore the reality of the people who live here. It's easy not to realize that, to stay in the tourist corners – especially since tourists here are afraid of everything, so they stay in their golden prison for a bit (Mike, Irish tourist).

I met Mike during a favela tour and interviewed him the next day. He pointed out a value of the favela that recalls the one already evoked: of the possibility that his visit offers for obtaining a profound knowledge, opposed here to the superficiality of a strictly beach form of tourism. Mike presented visiting the favela as a duty. Furthermore, he was not one of those who strived at all costs not to be assimilated to tourists, and on this point he had a lucid position:

> Obviously, we're tourists, but that doesn't stop us from learning things! Of course, we do all the touristy stuff too, but we don't limit ourselves to that. For me, two days on the beach is more than enough, and even for the kids, I want them to see something else. Otherwise, there was no need to travel so far.

On the one hand, Mike demonstrates a posture that could be considered typical of tourism, insofar as he insists on the importance of the most complete experience possible and shows, as far as the place visited is concerned, a search for exhaustiveness that, in a way, is included in the idea of a tour. The visit to the favela was thus perceived as important insofar as it enabled "doing it all" during his stay in Rio de Janeiro. The search for exhaustivity and the collector's attitude of the tourist who wants to "see everything" are, of course, among the driving forces behind the tourist promotion of the favelas, and are among the reasons that explain the development of favela tours. However, on the other hand, Mike maintains an opposition between the so-called tourist practices and the visit to the favelas. For Mike, including a favela tour in his stay was a way to make it more complete and to go beyond the superficiality of practices that he perceived as strictly touristy. But it is obvious that for others, going to a favela is a way to depart more clearly from the classic tourist circuit and to show some courage.

For some people, especially young backpackers, going to a favela is a practice that embodies a transgression capable of conferring symbolic prestige to the person who commits it. This is how I met young travelers in youth hostels, who themselves offered to accompany their clients to *baile funks*, popular favela parties. One Friday evening, I decided to accompany a group who chose to participate in this favela funk party[1]. The 10 or so clients of the hostel were waiting at the bar for a minivan, rented by the hostel, which would take them a little before midnight. The group was to be taken to the mythical funk club Castelo das Pedras, located in Rio das Pedras, a community in Jacarepaguá (western zone) whose classification as a favela is currently under debate. Beyond the fact that it is not located on a hill, the community has never been associated with drug trafficking, an important element in the popular classification. In fact, Rio das Pedras has been known since the 1980s as the birthplace of the militias, the paramilitary factions that hunt drug traffickers and take control of the area. Obviously, these nuances are not evoked by the hostel that sells a party in a favela and the music played in the club is so representative of the favela's popular culture so that no one questions their word anyway.

A good half hour after leaving the hostel and Copacabana, we arrived in front of the Castelo das Pedras. Although it was past midnight, the attendants, who were young employees of the hostel, said that it was too early to enter and directed the group to one of the *botequins* lining the sidewalk opposite the club. The bar, already, seemed to appear to everyone as an authentic place in the favela. The customers presented before our arrival, who seemed to be regulars, had their eyes glued to the television screen which broadcast a mixed martial arts fight, a sport which, as explained to us by the attendants, is very popular in Brazil, in particular in the most underprivileged layers of society. Visibly understanding what our conversation was about, one of the customers of the bar, leaning at the counter, turned to the group to affirm, in a form of Portuguese that none of the tourists understood, his love for the sport. Paola, one of the guides, translated it: "He says that all the world champions right now are Brazilian." To this, he added in broken English:

1 The replacement of the internationally known term "baile funk" with "favela party" in several of the city's hostels is also a sign of the emergence of the favela as an international brand and tourist attraction.

> Man at the bar: Yes, brasileiros! This one [pointing to the TV], carioca! Favela!
>
> Peter [a Dutch tourist]: Oh, does he come from a favela?
>
> Man at the bar: Yes, yes, favela!

The man, in high spirits, approached Peter to slap his hand, then embraced him. He started talking to him but I was unable to clearly distinguish the words, and Peter was unable to understand him, obviously. Next door, the other members of the group were sharing a beer and discussing their trip with the guides. As everyone had paid for the evening at the hostel, the attendants paid for the beers and, a few minutes later, paid for the entrance to the club.

The club looked like a gigantic hangar, very wide and high-ceilinged, lined with walls of speakers. With the exception of a few young party-goers, most of the group complained about the volume. Antoine, a young Frenchman, placed cigarettes in his ears:

> It's so loud. This is like the sound system of a U2 concert at the Stade de France in a room with 200 people. You can tell there's no regulations here.

The large group danced in a circle, obviously self-conscious, among the Brazilians who, for the most part, were dancing in pairs. Brazilian funk refers, in its lyrics as well as in its dance form, in a very explicit way to sexual acts. A few meters away from our little group, other participants seemed to flirt with the guides. After an hour or two, a large part of the group expressed the desire to return to the hostel. The guides, however, had fixed a departure an hour later, so that everyone could return together. Four or five of us decided to leave by bus anyway. Around 3 in the morning, the favela seemed less festive, and some of my companions for the night seemed to show a certain anxiety. When, to reassure them as much as possible, I asked passersby for a precise itinerary, some tongues were loosened: "It was special, anyway. I didn't expect that. I am completely deaf. It's really 'favela'," said Florent, a young Frenchman who was starting his university exchange year in Rio. While waiting for the bus that would take us on the road to Copacabana, James, an English tourist, pointed with his eyes to the electric cables above us: "Wow, there is really everything here," he said laughing.

It was the next day, at breakfast, that I met with those who, the night before, had stayed longer than us at the Castelo das Pedras club. Quickly, after having exchanged some impressions on our decision to leave, Marco, a Dutch tourist touring the world, told me about the end of the evening:

Marco: Well, we left around 4 in the morning. One moment, we heard a big boom. There was a guy rapping next to the DJ; he stopped right away.

Me: What do you think it was?

Marco: Well, a gunshot. You should have seen the DJ, the face he made, then there were people running. So we said, "Let's go home."

Me: Were you scared?

Marco: A little bit at first, but then there were also a lot of people who seemed to think it was completely normal, so it didn't last too long. This is Rio, man!

Beyond the pleasures of the party, it was certain that the young tourists took from this participation a means to value their way of discovering the favela. "It is much better than a visit; we participate in the real life of the favela there," declared Woulter, another Dutch tourist. It was certain, for the witnesses, surely more frightened than they would like to have admitted to at the time, that the gunshot became, the next day, the highlight of the stories of the night before, as if to impress those who had decided not to follow the group and had stayed at the hostel. As for those who had left earlier, the disappointment of having missed this event was palpable, as it was clear that this gunshot brought an ultimate touch of authenticity to the experience of the previous day.

In this example, it is evident that the experience was valued for its transgressive nature, vis-à-vis the tourist circuit being perceived as traditional. Going to a party was seen as an original way to enter a favela by these young people who, on the other hand, were very critical of favela tours, which could do little to satisfy their quest for original experiences. One of the goals of this narrative was undoubtedly to give themselves the dimensions of a real traveler, braving the dangers and the fears of many tourists to have an authentic experience. It is certain that, in these operations of valorizing the experience, the mode of discovery is sometimes as important as the object and that, in such or such environment or social group, all practices are not perceived as equivalent.

In any case, it is undeniable that visiting the favela is a valuable experience back in the country of origin. Indeed, favela tours do not only develop by way of a valorization of the favelas, but also by the symbolic value of the visit which is, in itself, likely to be attached to the traveler and to their practice. In other words, favela tours are also what Marc Augé calls a "visit to the future past" (Augé 1997). Motivated in part by the promise of being able to later say, "I've been there," these experiences add to the traveler's symbolic capital. So much for the positive aspect of the visits, without which there would surely be no tourism in the favelas. But the practice of favela tours is as much an opportunity to enhance the self and the trip as it is the risk of being exposed to critics. It is this double symbolic dimension that constitutes the sign of a competition.

4.1.2. *A criticized practice*

While participation in a favela tour may be valorized, it is also highly criticized and therefore carries a risk. By far the most criticized form of a favela tour is the classic, mainly motorized, tour. Companies like Favela Tour are regularly accused of organizing voyeuristic tourism, a "human safari" that turns the favela inhabitants into zoo attractions observed from the window of the guide's van. This criticism is very frequent among those who choose another form of visit, especially pedestrian and organized by the locals. Bianca Freire-Medeiros came to the same conclusions and stated that everyone is convinced of the superiority of their mode of visiting (Freire-Medeiros 2009). During my observations, I have often gathered this type of reflection. An event observed and noted during a local tour shows how participants place themselves in opposition to other tourists. During the tour, we came across a tourist minibus, whose presence was indicated to us by our guide. There followed a discussion between the participants of our excursion, who showed their disapproval of this other form of favela tour. We were a small group of four: an English couple, a Spanish tourist and me. Patricia, a Spaniard who had lived in Rio de Janeiro for three years without ever entering a favela, was the first to speak up when she saw the minibus:

> Patricia: It's very strange, though, isn't it? Why would you do that, when you can come with people who tell you about their lives from the inside? I don't get it.
>
> Paul [turning to our guide]: How do people here see it?
>
> Mario, Rocinha guide: Very bad. Put yourself in their shoes, imagine that someone has come to your neighborhood by bus to take pictures of you through its windows!
>
> Patricia: It's crazy, outrageous!

In this case, the criticism was encouraged by the guide, but it was not necessary for a condemnation of the favela tours to be expressed by the practitioners of local tours. This condemnation was based on two closely related aspects. On the one hand, the criticism had of course an ethical dimension. What the clients of the motorized favela tours are reproached for is, in a way, withdrawing from the gaze of the inhabitants in order to take a better look[2], and thus being guilty of voyeurism. On the other hand, they are criticized for their superficiality and mocked for their presumed conviction to discover the favela, while it remains inaccessible for them since they do not experience it on a sensory level. This is what Patricia expressed later in the tour:

> For me, it's important to feel, to walk the streets. For example, here it's hard [we are in the middle of a climb]. Well, that's part of the favela, too. How can you see that by bus [laughs]?

In both cases, it was the mode of visiting that was criticized. It is easy to understand the role that our own practices can play in devaluing that of others, despite the fact that many people have no idea of the different practices available, and either take part in a tour booked by their hotel, or follow the recommendations of written guides (*Lonely Planet* and *Guide du Routard* only refer to tours organized by non-local agencies). But the reference to motorcycle tours is also frequent among those who criticize outside the space of the favela tourist visits.

Some criticize tourism in the favelas, whether it is external and motorized or local and pedestrian. The simple fact of wanting to visit the favela is criticized since it testifies to a voyeuristic desire, the marks of which are the brevity of the visit and the absence of building a relationship with the inhabitants. The use of a guide is therefore, in itself, condemnable. This type of criticism can be found on the Internet, particularly on travel forums that raise the question of the validity of participating in a favela tour:

> But come on! "For or against" favela visits… The fact of paying a guide to accompany you to "visit a favela"… seems to me rather shameful. The favela is not a zoo! Walk around it, as you would simply walk around the streets of any city you visit, why not (no need to take out the camera, wear branded shoes or shout loudly that you are a foreigner), but at your own risk[3].

Here, the use of a guide is enough to condemn the practice and the idea that the emergence of a tourist framework for visiting the favela is enough to transform it into a zoo. The author of this message posted on the Internet therefore advocates a

2 See Chapter 8.

3 Translated from an Internet forum, see: https://voyageforum.com/discussion/contre-visite-favela-bresil-d298037-2/.

solitary visit to the favela, without using the services of a guide. Other forms of criticism insist instead on the need to be invited into the favela and the need to get to know its inhabitants:

> Personally, I find it disgusting. I go to these neighborhoods when I have made friends who, after a certain time, invite me to go there (and I'm careful, I'm not the naive gringo to be robbed). Otherwise, to go and see it up close, no thank you: it's the people who interest me, not the metal sheets or the rubble[4].

Critics, however, usually focus on classic tours and are less categorical in their condemnation of the mere possibility of favela tourism. Motorized favela tours have historically constituted a first mode of tourism in the favela, and undoubtedly a necessary step that continues to reassure some tourists. The financial redistribution they bring about is partial and questionable, especially since (as Chapter 3 has shown) it widely depends on the desire of the guides, who have the prerogative of tourist mediation in the favelas. Today, faced with competition from local tourism (which also reproduces these biases in the redistribution of profits), favela tours organized by non-local companies have largely declined and are probably destined to disappear. It is nevertheless flagrant that the criticisms, justified in substance, are formally developed in such a way that the tourists are always the blamed actors. This is an overestimation of the influence of demand compared to offer, and of the level of knowledge of the offer that individuals may have. Indeed, for the most part, tourists decide overnight to take part in a favela tour, responding positively to a proposal made by other intermediaries, according to the arrangements that the latter have made with guides or companies wishing to create a clientele destined for the favelas. Of course, tourists can be blamed for a lack of effort and awareness. However, it should also be borne in mind that, as far as the favela tourist offer is concerned, the boundaries between local and non-local are blurred and that, aware of the controversy they generate, a certain number of individuals from outside the favela but competent to transform themselves into mediators add an indigenous touch to their visits and endeavor to reassure the tourists about the redistribution of their profits. What is interesting and necessary to question therefore is not so much the existence of a critique as its constant focus on tourists. Moreover, considering, as I do, that the valorization of tourist consumption (even though the word "tourist" is refused by some) precedes its condemnation, we have every reason to believe that at least part of the criticism must be understood, within the space shared by those who condemn and those who are condemned, as the sign of a symbolic struggle for access to the positive symbolic charges that the favela carries.

---

4 Translated extract from the *Guide du Routard* forum website: https://www.routard.com/forum_message/3063547/visiter_les_bidonvilles_de_rio_de_janeiro_.htm.

4.1.3. *Anti-tourist tourists, between legitimization and criticism*

Criticisms of the favela tours are primarily directed at tourists. Accused of being voyeuristic, it is their desire to observe the poverty that is questioned, much more so than the companies that organize tourism in the favelas in the first place. To understand this, it is surely necessary to realize that the criticisms rarely take place outside this phenomenon. Most often, the most virulent criticisms are made by individuals who have another tourism practice in the favelas, even though they do not qualify their practice as being touristic. However, all the practices stem from a desire to discover the favela, relying on its valorization and giving rise to a temporary incursion into a favela, motivated solely by the desire of the person who goes there, and then sooner or later leaves. This is why, without totally relativizing the ethical dimensions of a tourism that indeed poses a certain number of problems, concerning mainly the redistribution of its profits, I propose to analyze the criticisms by insisting on the way in which they are personified in individuals who, from my point of view, share with those they criticize the same social and symbolic space. Indeed, a quick examination of the most frequent criticisms reveals a crucial fact: the condemnation of favela tours, whatever the mode in which it develops and the aspects it chooses to question, does not go hand in hand with a total abstention from a tourist practice within a favela.

I have already mentioned some examples of criticisms made by other tourists who have chosen other ways of discovering the favela: against motorized tours in the case of choosing a local tour, and against all guided tours in the case of individuals with connections in the favelas. It is worth mentioning another configuration, both more common and more surprising: participants in favela tours, including motorized tours, are themselves critical of the practice. Indeed, many tourists have told me that they have a very negative view of the existence of tourism in the favelas, all while participating in it. We find here – as in the other configurations mentioned, provided we agree on a broad definition of tourism – a paradox widely commented on by Jean-Didier Urbain, who evokes the omnipresence of the figure of the "anti-tourist tourist" (Urbain 1991).

University exchange students, for example, are a group who are particular bearers of this paradox label. They are a perfect example in that, from their point of view, the favela tours are almost unanimously perceived as "human safaris", "tourist stuff", whereas it is obvious that the favela nevertheless confers a form of attraction on them. Most of those I met had themselves enjoyed a tourist experience in the favela, either through "friends of friends" or by staying there for a while. Indeed, during the 2010s, more and more students started to rent rooms in the favelas (almost always in Vidigal as well as sometimes in Morro da Babilônia, therefore always in the South Zone), allowing them to access cheaper housing, and sometimes fulfilling their desire for a more "authentic" and more local experience. But what is even more striking is to observe that some of these students also participated in favela tours. By meeting these students during the tours, I observed a typical aspect of the tourist polemic that is built on criticisms always addressed to an imagined Other and that, in no case, forbids the practice for ourselves. The only condition is that we must be able to legitimize our practices and to distance ourselves from the imagined profile of those who would be the typical participants of favela tours:

> I find it a little too restricting as a visit, the tourists who come to see the poverty and everything. I do it because I've read a lot about it. It's really a political approach (Aurélien, university exchange student and Favela Tour tourist, July 2015).

The modes of legitimation used by students are primarily based on the objective conditions of their stay. They are not tourists, since the general reason for their mobility is not touristic. In this, they form a group very close to expatriate workers, who are also carriers of this paradox. For these two groups, it seems that the fact of being in a non-touristy mobility completely prohibits them to think of themselves, even occasionally, as tourists. This is, moreover, what constitutes one of the main motivations for the construction of an imagined tourism: to be a tourist would be a permanent condition. However, it is much more correct to think that we are tourists by circumstance, and that we indulge in tourist practices at times and in places, even though our mobility is defined differently and that we can be more permanently defined as business travelers, journalists, ethnologists, students, expatriate engineers, etc. But it is obvious that the application to distinguish ourselves from the figure of the tourist grows as it appears necessary to legitimize ourselves differently. Students and expatriate workers place a large part of the success of their mobility on the fact of having integrated themselves locally and having, through duration and involvement, precisely reached an experience inaccessible to tourists. Similarly, for journalists and academics, it is important not to be assimilated to tourists and to justify their status as mobility professionals. Everything, ultimately, is relative to the issues associated with mobility. Thus, backpackers are another example of individuals who have every interest in standing out from tourists, in their eyes as well as in those of others. Often, these positions lead individuals to choose other practices. This is what I comment on in this chapter. However, it is interesting to note that, in some cases, they simply participate in favela tours, without this participation being assimilated in their minds to tourism.

4.1.4. *The denial of the tourist setting*

The valorizations of the favela experience, whether or not they are supported by the criticism of other forms of experiences, have in common that they operate through a denial of the tourist character of the experience. Indeed, in the conversation transcribed at the beginning of this chapter, Marc omitted to recount the conditions of his visit to Rocinha and mentioned neither guides nor other tourists. I have observed this phenomenon on many occasions, and it is sometimes only through my knowledge of the content of the tours, acquired through repeated observations, that I have been able to detect the sign of the omitted tourist setting in the account given to me of a favela experience. In the cases I was confronted with, after having asked the person about the conditions of their experience, the admission was immediate and without shame. The tourist setting was omitted from the story, but the omission did not seem to turn into a lie, probably at the risk of being found out. In the case of the students with international mobility, for example, the reason for their presence offered a justification in itself: not being tourists, they could engage in the same practice, convinced of the superiority of their perspective. Moreover, the reasons for their visit (to get to know the depths of a city they were going to live in for the next few months) undoubtedly seemed better to them than those they associated with "tourists": voyeurism, morbid desire, sensationalism, etc.

In any case, the omission quite successfully revealed the ambivalence at stake here, between valorization of the favela experience and the risk of devalorization if in case of an assimilation to a tourist practice. It also shows the gap that can exist between the lived experience and its narrative, and the necessity to transform the experience by storytelling to preserve possible symbolic profits. Finally, it reveals the difficulty in defining ourselves as a tourist and the concomitant tendency to classify the other, even a disembodied other that is more ideal than real, in a "tourist" category created to better distinguish ourselves. Thus, the individuals observed in this survey were to a large extent both tourists and anti-tourists, and this paradox seems to play a key role in the evolution of tourist practices, insofar as they work simultaneously to the dissemination of criticism transformed into norms, and then to the diversification of practices.

I have observed the post-tour concealment of the tourist framework in a sufficiently common way to be able to affirm that it is more related to a certain type of visit than to another. In fact, it concerns primarily tourists who have participated in the most classic form of favela tours, that is, a guided tour under the self-guidance of a non-local company like Favela Tour. This is not, in most cases, a lie, but rather a clever omission of the positioning of the tour within a context. Without directly affirming it, these accounts leave room for the narrator to be pictured as if they were alone in the favela. An almost opposite phenomenon is observed in the case of participation in local tours. In these cases, the narratives include numerous amounts of information about the guide's personality and history. For example:

> I visited it [Rocinha] with a local guide. He was born in the favela and he knows everyone, so it was great. We had a true point of view... of the inside. He told us how things really were in the favela. And what

> was good was that he didn't just talk about the problems, he talked about everyday life (Sarah, American academic during an informal discussion in the United States, April 2016).

4.2. Distinction, the driving force behind tourism in the favelas?

Gathered in a common analysis, like two sides of the same coin, valorization and devaluation of the tourist experience of the favelas are presented as driving forces for the development of favela tours and the diversification of their form. The favela appears as a space with a symbolic charge activated by the visit. But, as we have seen, the way in which this is done can be of great importance depending on the actors involved. Indeed, it is sometimes necessary to be able to distinguish our practice from that of others. I have said that this demarcation has two main sources. On the one hand, we try to distinguish ourselves by choosing a mode of experience that is different from another that is simultaneously described negatively. In order for a practice to be perceived as having a superior value, one or several others must necessarily be criticized at the same time, whether this criticism is already established on a large scale and inscribed as a norm of "good tourism" or whether they participate in developing it themselves. On the other hand, I wanted to insist on a fact that, due to the lack of attention paid to the stories collected far from the field, we have more often than not the tendency to ignore them. This fact consists of a gap, sometimes large, between the real practice and the way it is told. It highlights the existence of an accumulation of practices and their critics, which this time is essentially a matter of language. In fact, I have observed many cases of people who, although they have a point of view as acerbic about the negative aspects of the favela tours as those who choose another form of visit, still participate in them. In these cases, it is a legitimation a posteriori of the participation that operates the demarcation on the motive of the perspective, which is most often considered to be lacking in the others. Finally, linked to this imperative of legitimization, there is also a concealment, in the form of omission, of the tourist nature of the experience in order to restore in the narrative only the point of contact between themselves and the favela. I will deal with this phenomenon at the end of the chapter since, in order to understand it, it is necessary to set up a theoretical framework inspired by Pierre Bourdieu's studies on the distinction. For the moment, by focusing on the choice of alternative practices, I would like to underline the driving role played by the valorization/critical pair for the diversification of the tourist practices in the favela. This question, which will lead us to talk about distinction in Bourdieusian sociology, is based on the thinking of Jean-Didier Urbain, an indispensable reference for what concerns questions of valuation and devaluation of tourists and their experiences.

4.2.1. *Ritualization and de-ritualization of tourism practices*

Jean-Didier Urbain (Urbain et al. 2008) proposes to divide into three categories the different choices made in terms of visiting a place, taking the example of the museum: an initial practice, an experimental practice and an interstitial practice[5]. The visit to the museum under classic conditions constitutes the initial practice. The experimental practice, developed by those for whom the classic conditions appear too poor in benefits, or even potentially harmful to their reputation, consists, for example, of visiting the museum in the more private conditions of a private viewing. Finally, others adopt an even more extreme posture and choose to visit the museum illegally, by being locked up at night: this is the interstitial practice.

	Initial practice	**Experimental practice**	**Interstitial practice**
Visit to a museum according to the example of Urbain	Visiting the museum during opening hours	Participating in an exhibition opening	Being locked up at night in the museum
Nature of the visited space	Open to all	Open to some, closed to others	Closed
Nature of the visitor	Customer	Guest	Intruder
Nature of the presence of the visitor	Standardized, expected and collective	Standardized, selected and exceptional	Not standardized, unplanned, unusual or even prohibited
Nature of the visiting experience	Standardized	Customized	Transgressive
Practice in the favelas	Favela tours, guided tours during the day	Events and experience, accompanied by a local	Independent visit, without the use of a guide and without invitation from a local. Visit at a time when the doors to the favela are closed

Table 4.1. *Typology of tourist practices in the favelas*

5 In *L'Idiot du voyage* (1991), Jean-Didier Urbain instead divides tourist practices into two categories: initial and interstitial. The interstitial category is itself divided into two, with "secrecy" practices, on the one hand, and "risk" practices, on the other. See Urbain (1991, p. 303). I take up here a typology evoked in an interview between Urbain, Anne Doquet and Olivier Evrard, and accompanied by the example of the museum commented on later.

This brief typology of the modes of visit constitutes a heuristic model that it is a question of confronting the case of tourism in the favelas, with the aim of ordering the diversity of the practices that are linked to it. Rather than arriving at a typology, the aim of this approach is above all to show that the different practices are indeed mutually related, and that together they form a system within which tourists situate their practices. To pass from the example of the museum to that of the favelas, it is advisable to widen the typology of Urbain and to bring about other implications. These implications are listed in Table 4.1, which shows, according to the three categories he put forward, the differences in the modes of the visit and their consequences on the nature of the experience lived by the visitors.

Moving from the example of the museum to the case of tourism in the favelas, the model undergoes some significant modifications. Indeed, the two spaces are not of the same nature, especially in terms of their "openness", since a museum opens and closes its doors, whereas the favela is a public space that, by definition, is always open. However, this openness is relative in the eyes of tourists. It should even be considered that, for decades, before the tourist presence was somewhat normalized in the favelas, they were considered inaccessible, dangerous spaces and therefore closed to visitors. By going from the example of the museum to that of the favela, we move from a case of institutionalized access to a case where the barriers are symbolic and where the guide and/or the inhabitant are the key persons, insofar as they enable passing through an invisible door that, in their absence, remains largely perceived as closed. It is worth mentioning that favelas can be completely closed to tourists, especially when police interventions or gang wars take place there. In this case, the tourist presence has more explicit value as an interstitial practice, because interstitial practices are always linked to risk-taking. While I cannot attest to such practices in relation to a day trip, several long-distance travelers have experienced such moments, when they had settled in a favela for several months. Although they do not fall into the category of "visitors", it is nevertheless true that their experience of the favela, precisely because it has blurred the distinction between the tourist and inhabitant (and more often than not denies the tourist nature of the experience), has a higher symbolic value when compared to the initial practices. The same is true for experimental practices situated between an objectively tourist situation and a quasi-residential situation that completely denies its tourist nature. Here, the range of practices is wider, and these differ from the others simply because the visitors can distinguish themselves from a classical visiting condition, but only towards the participants of the initial practices, and not towards those of an interstitial practice who, for their part, would certainly not make a strict difference between the different guided tours. I have placed in the liminal category practices that may, for some, also appear in the other two categories. At the most "initial" extreme of practices, I mention those that differ less from the classic favela tours in their objective form than in the way they are experienced by visitors.

Tours organized by locals are one of these. Formally, they differ only slightly from the favela tours organized in the former. However, the presence of a guide who is, or presents themself as, a local can transform the visitor into a guest. There is, of course, a commercial relationship, but it is framed so that it only takes a minimal place during the visit, and then in the tourists' narrative. In the case of this type of visit, the emphasis is placed on the experiential dimension: we enter the houses, we sometimes play soccer with the children, we discover the trades of the favela, etc. At the most interstitial extreme of the experimental practices, there are practices that contrast in a more flagrant way with the favela tours: invitation by a local in a non-market relationship, participation in a party (*baile funk* type). The difference here with the guided tours is based on two things potentially accumulated: having something to do in the favela (being there for a specific event) and being invited. In general, this table is not meant to be fixed; it intends to model a categorization of tourist practices in the favelas, which the individuals who visit the favela use to differentiate their practices. None of them has the same exact perception or use. Also, visiting the favela with an inhabitant can be perceived as an initial practice by some and experimental by others. In the same way, taking part in a classic favela tour can at times, especially in the narrative, take on the appearance of an experimental practice.

The advantage of using this model is to restore to tourism in the margins its innovative nature and its structural functioning, once it is taken into the whole of the tourist practices. To continue this reflection, I would also like to try to show that this typology can also be used not in a strictly synchronic way, but with a diachronic dimension for which the case of the favela tours appears to be more telling than that of the museums. Indeed, it is possible to consider the history of tourism in the favelas in this way, insofar as we can consider that there was a time when only the initial practice (the classic favela tours) existed, and that this in itself had a positive symbolic charge that had not yet been attenuated by the existence of alternative practices. The aim is to show how practices tend to differentiate themselves from an initial (reference) situation towards an interstitial situation, under the impulse of the strategies employed by tourists to demarcate their practices, in order to reintroduce the feeling of innovation and exploration. Jean-Didier Urbain calls this strategy "de-ritualization". In fact, according to him, initial tourism is similar to a ritual, collective practice of commemoration, celebration, recollection, etc. It moves around sacred places which, together, form a system, a code (MacCannell 1976). It is from this ritualized form of tourism that dissident practices develop, forming a "space of contestation, even unconscious, of the obligations prescribed by ritual tourism" (Urbain 1991, pp. 301–302). The innovative practices are thus only in reference to a tourism that serves as a model. However, they can themselves be integrated into it and become a model towards which tourists wishing to distinguish themselves, and thus refusing their status as tourists, will have to distinguish themselves. As

Urbain writes, "de-ritualization becomes fashion, then fashion becomes ritual" (ibid., p. 308).

First, a diachronic consideration of the initial/experimental/interstitial model leads to the relation of all of the tourist practices in the city of Rio de Janeiro. In this case, the initial situation is that of a tourism focused mainly on the recognition of heritage, the visiting of monuments, beaches, etc. This is classic tourism in Rio de Janeiro. The de-ritualization of this situation, which in large part can be understood as a search for authenticity of both our own experiences and its object, leads to a more experimental practice. Here, the tourist is looking for a more personalized experience, an encounter with the "real Rio", and therefore practices that focus more on meeting local people and participation than on the visit that places them in the position of a spectator. Finally, the emergence of tourism in the favelas can be interpreted as a result of a second de-ritualization that takes the first two types of practices as references. Here, the authenticity of the object is assured by its hidden, unknown, non-touristy dimension, etc. We are at the heart of the negotiation of the touristic paradox, that is, practicing tourism in a place where tourism is perceived either as impossible or as simply absent in comparison to the rest of the city.

In a second step, we can question in the diachrony the evolution of the tourist practices in the favelas, qualified as interstitial practices on the scale of the city. In fact, if we look at the scale of the specific space, the favela, we find quite easily the same structure and the same structuring forces. Thus, the favela tours, located at the interstitial extreme on the scale of the city, ritualize themselves and become the starting point (thus the initial extreme) of a new enterprise of de-ritualization, and thus of innovation. In this perspective, the initial practice is represented by the favela tours, grouped visits (10 people) of a maximum of half a day, that is, the reference model that will then be de-ritualized. The experimental practice thus takes multiple forms: we turn to more participative visits, such as the *baile funks* I described above, as well as a certain number of visits where the tourist is accompanied or not by a guide who, this time, is a local who can therefore hide their profession under their identity of favela inhabitant. Finally, the interstitial practice is still extremely rare here, and I only had rare occasions to notice its existence. It consists, for example, of going to the most dangerous favelas, where the initial tourism has not marked out the modalities of the encounter between the foreigner and the favela.

This typological reflection, inspired by Urbain, seems to me to be useful in the approach to this subject, despite the multiple simplifications it entails. Within and between each category, there are hundreds and thousands of stories, nuances, not to mention everything that should straddle the categories. In reality, this schema is constructed on the scale of practices, not actors – in which case linearity should be abandoned. In no case can we consider that the actors all have in mind a structure like this, nor that there can be any shared structure. It would almost be necessary, for each individual, to draw a specific diagram, a sort of mental mapping of potential practices, where what appears classic and initial for some is interstitial for others. However, this idea is well understood by tour operators, and the schema validates Urbain's idea quite well when he states that tour operators "camp on the border of initial tourism and experimental tourism" (ibid., p. 306). Indeed, in the case of favela tours, it is important for tour operators to maintain their place on both scales. This is evident in the two main efforts of tour operators to, on the one hand, maintain the marginality of the object, by insisting on the favela as a mystery unknown to tourists as well as to all those who do not live there, for example, or by exaggerating the necessity of the guide, while, on the other hand, part of their work consists of taming the marginality for the tourist, reassuring them both ethically and on the security level.

The interest of this approach is to propose an alternative, which I find effective, to a set of interpretations on tourism development. It allows, for example, us to nuance the accusations of voyeurism and to include them in a structure. For it is a real structure of tourist practices that we are talking about here, insofar as a large part of the meaning of practices can be understood by their relative positions in a whole. The construction and evolution of practices can be understood as the results of a permanent negotiation between ritualizing forces (which, in a way, tend to create heritage) and de-ritualizing forces, which can be largely assimilated to strategies of distinction, and which grossly tend to create margins or push tourism towards the margin. In short, it seems to me that this makes it possible to extend the questioning triggered by Urbain and more specifically to show how interstitial tourism and the tourist taste for the margin can be explained in a systemic way.

4.2.2. *Distinction in tourism*

What my approach to the diversification of tourist practices in the favela and my use of Urbain's typology imply is that these practices are interspersed with issues of distinction that structure a competition for access to the positive symbolic charge that the favela and its visit represent. This also presupposes the existence of a social space of leisure mobilities that enacts norms and values, and divides between good and bad tourists or, more exactly, between travelers and tourists, the figure of the tourist being always negative from this point of view.

The concept of distinction fits in continuity with the works of Pierre Bourdieu in *Distinction: A Social Critique of the Judgement of Taste* (Bourdieu 1984), where the sociologist exposes a theory of cultural legitimation. In Britain in the 1970s, Bourdieu showed the correspondences between the hierarchy of social groups and that of cultural practices. He showed how the elites dictated the conditions of legitimization and valorization in matters of culture, dominating the processes for classifying good taste and legitimate culture. From a hierarchy of the forms of cultural consumption, the individuals and the groups delivered themselves a fight to build, relatively to this hierarchy, their cultural capital. The cultural capital, that Bourdieu added (in addition to the concepts of symbolic capital and social capital) to the economic capital of which Marx had made the center of the relations between classes, is another system by which the agents modify or consolidate their position in the social space. Indeed, cultural capital is a set of resources that can be converted into other forms of capital, but it is also necessary for its own accumulation, for if economic capital can be transformed into cultural capital (the purchase of a work of art, for example), one aspect of what constitutes cultural capital consists precisely of knowing legitimate culture and mastering its codes.

In Bourdieu's view, legitimate culture is dictated by the elites, but all agents have a sense of it, the middle classes being roughly preoccupied by a double duty of distinction from the working classes and imitation of the upper classes, while the working classes are characterized by "the choice of the necessary" excluding themselves from legitimate culture without questioning it. Distinction is a constantly renewed process, as other social classes imitate the upper classes and as the latter, in reaction, reinvent other practices that will be imitated again, causing the process to repeat perpetually. The agents and the groups thus fight for the acquisition of a place of choice in the social space which, contrary to the Marxist theory, is not summarized to the appropriation of the means of production, and thus to the accumulation of the economic capital. The social space is constituted as an effect of these struggles:

> Agents and social groups are distributed in a theoretical space of distinct and hierarchical positions, which are related to each other and distinguished in terms of partially competing principles of differentiation (Duval 2020, p. 306).

Indeed, the means of distinguishing ourselves are numerous, and the possession of significant economic capital is only one of these means. In *Distinction*, Bourdieu endeavored to show how cultural practices have a dual relationship with the structuring of social space. Indeed, the choices made are at the same time determined by the social position and determine it in return. This is the meaning of Bourdieu's formula which describes individuals as "social subjects classified by their classifications". In the field of tourism, in the choice of destinations, objects and practices, we find this double dimension: the position in the social space influences the choice of practice, and in return the practice allows us to reinforce or increase our capital, and thus our place in the social space. In Bourdieu's sociology, the capital that situates the individual or the group in the social space has four norms: economic, cultural, social and symbolic. The structural effects of economic capital, that is, of the economic and cultural resources, were known before Bourdieu's work, and it is especially the addition of other forms of capital which emphasized Bourdieu's contribution, in particular with regard to Marx. Cultural capital is the set of cultural resources, know-how, skills, qualities of expression, intellectual references, etc. Social capital consists of the social relations that an individual has. Finally, symbolic capital is all that can offer an individual a particular social recognition.

These different divisions of capital maintain, in practice, close interdependent relationships. First, they are mutually convertible. Indeed, the term "capital" has the advantage of showing that what is at stake concerns both the possibility of investing in this capital to improve the positioning in the social space, and that of enjoying the positioning to increase the volume of this capital. To emphasize, as Bourdieu does, that there are different types of capital is to show that one of the ways of investing in one of these capitals can be to convert it into another in order to better serve the agent's interests. Thus, economic capital can be invested in order to increase cultural capital, through access to the best schools, through the acquisition of works of art, etc. It seems that the position in the social space is improved precisely by these operations or circumstances, during which forms of capital are invested in to be transformed into others. The mechanism of reproduction supposes that all forms of capital benefit the others. Second, it seems that symbolic capital, which is central to the distinctive tourist/traveler dichotomy, has a special status. Certainly, it is itself convertible to the acquisition of all other forms of capital. Indeed, according to Bourdieu, symbolic capital functions as a "credit" understood in the literal sense, that is, as an advance granted by others on the basis of their social prestige. In this, it allows an obvious conversion. However, it is equally true to say that all other forms of capital are converted into symbolic capital, so that it sometimes seems difficult to make symbolic capital into a capital of equivalent status to the others. Pierre Bourdieu himself referred to this ambiguous nature of symbolic capital, defining it ultimately as economic, cultural or social capital, but in its perceived state. This does not mean that symbolic capital is less important than the others. Quite the opposite, it holds an essential place and could well be defined as what social agents ultimately seek. Bourdieu eventually indicated that "it might be better to speak, in rigorous terms, of the symbolic effects of capital" (Bourdieu 2000, p. 242). Thus, the possession of strong social, economic or cultural capital is a source of symbolic capital, and this symbolic capital serves to accumulate power and to improve the perception that others have of my position in the social space. Thus, it can be said both that nothing is only symbolic and that everything is symbolic. However, the practices and goods that we are talking about when we speak of the symbolic economy have precisely the characteristic of being purely symbolic and entirely disconnected from any strategy for improving the positioning of ourselves in social space. Indeed, what Bourdieu calls symbolic economy is based on a principle of denial of market aspects and interests: to attract prestige, agents must be able to present the acquisition of symbolic goods as unselfish. This is the case of tourism, which, as a cultural practice with strategies of distinction, functions on the model of symbolic economy.

Historians trace tourism as a cultural practice back to the 18th century and the development of a collective leisure mobility of young aristocrats traveling from Britain to Italy (Boyer 1999). This new travel practice, called the Grand Tour, differed from other ancestors of our leisure practices, in particular from the Roman *otium* and the *villegiatura*. It differed from it first of all by taking the form of a tour whose course was more or less strictly delimited, and which made it possible to circulate through the great sites of antiquity and of European aristocracy. The first distinctive feature of the Grand Tour was its quality as a training trip. Indeed, the young British and then European aristocrats completed their education, their mastery of high society mores, and their learning of languages (Wagner 2007). The Grand Tour was not a romantic wandering, but a journey turned towards the benefits of the return: Daniel Roche affirmed moreover the propaedeutic of the journey (Roche 2003). Certainly, it had an educational goal, but certain historians like Marc Boyer do not hesitate "to distinguish oneself was the essential motive of the Tour, more than the educational value attributed to the voyages" (Boyer 2002). For the historian, the development of the Tour corresponded to the emergence of a new way of traveling (quite distinct from migrations and pilgrimages), made of "idle practices", codified migrations and places of exception. It was through these inventions that the Tour distinguished itself from student mobilities, which had already appeared in the Middle Ages (Karady 1998), and prefigured tourism. Codified idleness was transmitted outside the Tour and, still in Great Britain, to the appearance of summer and winter resorts in the Midi French region, and then of sports. For Boyer, tourism is an "invention of distinction" which, in the same way as sports and fashion, does not respond to any other logic than that of distinction, that is, the renewal of social divisions and the superiority of the elites.

In various ways, leaving our homes for no other reason than for leisure was, in itself, a distinctive practice. This is how tourism was first interpreted in the light of Thorstein Veblen's *The Theory of the Leisure Class* (1899). For Veblen, the 19th century saw the emergence of a social class characterized by ostentatious consumption and the promotion of values that contrasted with those of the working classes. The leisure class valued the useless, praised the non-vital and the non-productive, overvalued leisure and devalued work. For them, the wandering of leisure mobility was the affirmation of a way of life and the demonstration of extraction, even though only temporary, from a life dominated by work. For Marc

Boyer, the elites "invented a counterculture of ostentatious idleness[6] that opposes them to the rich entrepreneurs" (Boyer 2002). This symbolism, which superimposes on the dichotomy of leisure/work a dyadic mobility/immobility, is still active. Going away is still a value in itself in a society like France, where 60% of individuals do not go on vacation. But, historically, the time of travel as an ostentatious practice corresponds to a stage that was exceeded as the leisure society expanded (still relatively). Indeed, it is to the extent that the mere fact of leaving became in itself less and less distinctive that the forms of mobility became more complex and diversified. The practices of leisure time were perfected and refined under the impulse of gatekeepers, themselves belonging to the elites and imitated by their class, always in search of new practices capable of distinguishing them as the old ones spread to the rest of the society. As far as leisure mobility is concerned, the search for places and the ability to partition these places, physically or symbolically, in order to preserve their enjoyment was part of the distinctive process. There is a geography of the most valued places of tourism, whose access is financially prohibitive, and long-distance travel remains inaccessible to a significant part of the population of the most traveled countries. But, in cases where access to places is no longer reserved, there is always the possibility of distinguishing ourselves by our uses, of staying there longer, of being able to open doors that remain closed to the majority, and even of getting there differently, formerly in luxury, today also slowly.

From the outset, we can therefore distinguish two levels on which leisure mobility has a distinctive dimension. The first concerns the practice of travel insofar as it is distinctive in itself. I will not say much more about the first distinctive dimension of travel – travel itself – because the case I am commenting on here, visiting favelas, is more specific. To distinguish ourselves by leaving is to distinguish ourselves from those who do not leave, in relationships and interactions to which I do not have access. The other, where our reflection leads, is related to differentiated practices and values that operate within the practices of leisure mobility. The visit to a favela is rather situated at the second level, that of a distinction by the ways of visiting, by the content of its leisure mobility. And this is itself made up of classifications and choices in a variety of fields that I propose to artificially separate into two categories. The first relates to destinations, places, cultures – in short, the objects of practice. The second relates to the ways of experiencing these objects, thus to the practices themselves, to their duration, their depth, etc., as I have discussed in this chapter (section 4.2.1) with the idea of a typology of practices, which remains too schematic.

First of all, it is obvious that not all destinations, not all objects of tourism, are equal and that some are more likely than others to generate a profit of distinction for those who go there. This is true from the time of the Grand Tour, which delineated

6 The concept of ostentatious idleness is here inherited from Veblen (1899).

sites of exception relayed in guidebooks, and it was also the case in the time that followed the Tour, when resorts started to pop up[7]. The emergence of tourism in the poorest parts of the cities belongs, however, to a different category of distinction. The value of places on the Tour depended on a collective narrative, which prefigured the logic of heritage in a mobility intended to recognize ancient heritage. It is already different in the case of the resorts, whose value is based on the privilege of being among ourselves, on the affirmation by the presence of belonging to the good society. The case of the favelas is different, but far from being isolated. On the contrary, it is even typical of what can be distinguished in modern tourism by the access to remote areas, excluded in principle from the mobilities of mass tourism. In this case, the symbolic value of the place is constructed within a mythology whose dominant aspect is the valorization of the traveler and the affirmation of the value of the exploration. Indeed, and as Urbain has shown, it is today on the side of the explorers, an extinct species, that we tend to look for the heroes of the journey. The favelas have, in themselves, a positive value in the mythology of the journey as exploration and adventure, as the expression "off the beaten track" has. I have already discussed this issue in the previous pages, showing the position of the favelas within the general practices in the city of Rio de Janeiro. Moreover, Chapter 5 will be largely devoted to it.

Second, and precisely because they stem from the mythology of adventure exploration, tourist practices in the favelas are interspersed by various operations of classification of these practices themselves. This is what I have shown by applying to tourist practices in the favelas the initial/liminal/interstitial. The distinction, the hope or the imperative to distinguish ourselves is also relative to the means of visiting the favela, and leads to a diversification of practices. However, the force of the classification operations remains relative because the favela remains a symbolic space inscribed in the rhetoric of exploration, at least until favela tours are considered as completely normal practices, and even banal. They are relative because, as I have shown by reporting discourses while concealing the tourist character, the modes of visiting can be transformed in their narration and the visits, even the most classic, can be presented from a new angle that restores their original value. The normalization (relative therefore, and concealed both by tourists who tell stories and guides who promote) of favela tours has begun to lead to a phenomenon of diversification. With the appearance, with greater intensity, of an ethical question accusing mass tourism, the promotion of tours organized by locals can also be interpreted in this sense. Other practices, those which consist of spending a long time in the favela, are immediately part of a relationship with the favela tours that reinforces the symbolic value of the former to the detriment of the latter.

---

7 Brighton, for example, was one of the first resorts favored by the British aristocracy, who abandoned it as other social classes arrived (Boyer 2000).

4.2.3. *The legitimate culture of travel*

In order for the Bourdieusian concept of "distinction" to be applied to it, it is necessary to define what the legitimate culture of travel is. Rather, there are legitimate cultures of travel, since tourism, in its broadest sense, covers a multitude of different practices that have not evolved according to the same codes, nor are they shaped by the same mythologies. In dealing with tourism in the favelas, it is appropriate to separate these practices from others, which it is commonly accepted to gather them together in the category of tourism and thus to abandon the idea of processing them together, as if they could fall under similar social norms. It is therefore necessary to specify to which category of tourism favela tours belong. Several terms have been used to describe this type of practice. The idea of slum tourism, which brings together researchers on the question of the tourism of urban interstices, seems to me to be too restrictive for this purpose, because the values and norms that apply to these practices go far beyond the question of the "slum", the idea of which must nevertheless be retained, since it designates a type of unusual space for tourism. The idea of poverty tourism, or poorism, is of the same type. I am more attracted to the idea of reality tourism, which is well suited to the practices in the favelas, through which tourists seek to experience a hidden reality and, according to the tautologies sometimes used, the true reality, the authentic reality, etc. Chapter 5, which deals with authenticity, is entirely devoted to this question. For the moment, what I want to point out is that the way of specifying the object of the visit – the reality – implies that the code of the "good experience" applied to the favelas is structured by a particular mythology that is not exclusive to it but has a particularly strong influence. This mythology is that of the traveler/explorer and their opposition to the tourist.

An excursion "off the beaten track", as the one in the favelas is presented, is in fact part of a rhetoric of valorizing travel, as opposed to tourism. The very choice to visit a favela can be a desire to escape from the condition as being a tourist – a paradoxical desire, insofar as the visitors then engage in the most organized (and often the only guided) practice of their stay – by crossing a symbolic border that has long kept tourists out of the favelas. We have seen the many ways in which the tourist character of the visit was either avoided by the choice of alternative practices, or denied in the course of the accounts given of the visit once back home. This is because the ideas of exploration and adventure that form the symbolic value of the practice are in direct contradiction with the normalization of this tourism, which gives the idea of a standardized product and of banal and non-distinctive practices. The ethical criticism is directed precisely at the "tourist" dimension, and is aimed mainly at those who resort to the "tour", symbolized by the minibus device. The avoidance of these devices and their concealment in the narrative allow the practice to regain its adventurous dimension. The opposition between the tourist and the traveler is old, and the term "tourist" has been pejorative almost since its appearance.

The tourist/traveler opposition is based on nothing more than the strategies of distinction employed in the mobility of Westerners[8]. These strategies exist within the world of leisure mobility, especially when its aim is to maintain the feeling of a unique exploration, and not a standardized tour to destinations that are already too well defined. But other forms of mobility, including professional mobility, also participate in the mythology of travel. As ideal types, investigative journalists, ethnologists, travel writers, etc., are valued figures of travel, who continue or replace the old figures: those of the explorer, the missionary and the adventurer. As real and embodied persons, these figures are themselves often part of processes of distinction. These professional travelers, instituted from the 19th century, participate in the devaluation of the tourist and in the very creation of this figure in relation to which they are insidiously at odds. More or less consciously, these types of travel actors, professionals and amateurs considered as models to be followed, organize their own distancing by devaluing the figure of the tourist and constructing codes that allow them to deny others the sharing of the same desires of discovery. They maintain the idea of an opposition between good, ethical and original practices (those of the traveler) and bad, imitative, consumerist practices (those of the tourist). In the case of the favela tours, I wanted to show, through the denials of the tourist character of the experience or the affirmation of the "backwardness", that this essential distinction is largely played out in the discourse, even though the practices can be identical. Here, the tourist does not need to be a real person whom we know, but can remain a scarecrow or an ideal foil for our own practices. These processes of distinction are sometimes essential for those who engage in a practice of exploring reality (a hidden reality, moreover, in the case of the favelas) and who intend to enhance their experience for themselves and others by showing their conformity to the legitimate culture of travel. The mythology of the traveler, an independent and reckless individual who goes where others do not, or in a different way, is also a euphemization of the class privileges that preside over the acquisition of the right values and skills demonstrated in the exercise of mobility. It inscribes in the symbolic order a competition for access to prestigious travel.

It remains to be seen how the distinction functions and runs through the world of travel, leading those who practice mobility to situate their discourses and practices in a symbolic order that opposes tourists and travelers. It is necessary, however, in the face of discourses that tend to give an apparent objectivity between what is "a difference of degree" (Urbain 1991) and not of nature, to situate the sets of practices

---

8 Examples elsewhere have shown that, like the question of authenticity as the central quest of tourism, the tourist/traveler distinction does not exist.

and discourses in a common universe and in a system within which they occupy relative positions. For Urbain:

> All engaged henceforth to diverse degrees in the ethnography of our modernity, traveler and tourist are registered in a vast movement of observation and recognition. They are carried by the same flow which feeds and reconstructs unceasingly our vision of the world. Tourism is not the degrading massification of travel. It is indeed the generalization of a mode of knowledge (ibid., p. 120).

4.2.4. *The functioning of the distinction*

The theory of distinction has been used and/or criticized many times, without always being well understood. Indeed, at first sight, it can give the impression of a permanent competition, of agents multiplying strategies to take advantage of their position. Of course, the Bourdieusian social world is made up of this type of strategy, but we probably focus too much on this positive and active aspect of the distinction. In other words, the actions that are most often focused on are the choices made by agents in order to put themselves in a valorizing position. Without being false or even rare, these actions are not dominant in the case of favela tours, and I do not think they could be in any other field. It is better, in fact, to shift the gaze to actions by which the agents engage not for profit, but above all for fear of being devalued. For, in each action, in each expression of taste, in each cultural choice, the state of our cultural capital is revealed and, as a result, we take a risk. It is highly probable that the set of risks incurred in the display of our cultural capital exceeds by far the cases of obvious and direct profits. This is the case in terms of occurrences in social life, and it is certainly the case in terms of individuals. Indeed, those for whom the expression of cultural capital is an easy act in which we engage as soon as possible, with the aim of obtaining a certain gain, are not the majority in the social space. For the majority, the risks of devaluation of the cultural capital are much higher and to work to demonstrate its value is not an act of maximization of its advantages, but of minimization of the risks incurred when we are placed in a situation which exposes cultural capital. This is why, moreover, the popular classes of the Bourdieusian model prefer not to engage in the legitimate culture whose codes they master poorly.

There is no doubt that too much focus has been placed on the elites in commenting on *Distinction*. This is of course due to the central role they play in the definition of legitimate culture, and to the fact that it is distinction, not devaluation, that gives the book its title. This fact is not to be criticized because, in Bourdieu's theory, it is indeed the distinction that is the driving force of the operations of classification, while the devaluation (of others) seems only a means to achieve distinction. But this emphasis on the historical process – distinction – has the effect of relegating to the background the concept without which it makes no sense: legitimate culture. Before classifying people in relation to each other, cultural tastes and practices are themselves classified and hierarchized. There are forms of culture whose appropriation is capable of valorizing and others that are not, some of which even lead to a devaluation. It is regarding defining the legitimate culture and its functions that the criticisms of the Bourdieusian theory are the most just. Among these criticisms, Dominique Pasquier has shown that the socialization of high school students was carried out less on the basis of legitimate culture than on the mastery of the codes of mass culture (Pasquier 2005). For her, Nathalie Heinich showed on the subject of contemporary art that there can be an "assumed rejection" of the culture of the elites (Heinich 1997), which does not consist any more of a self-exclusion but of a devaluation from below. It is obvious that the Bourdieusian model has suffered from the passage of time and the blurring of boundaries between legitimate culture, mass culture, alternative cultures and subcultures. Moreover, some researchers do not hesitate to say that the moment commented on by Bourdieu is in reality a very particular historical moment, qualified precisely by the creation of competition of economic capital and cultural capital in the establishment of the differences of classes (Di Maggio 1987; Coulangeon 2004). This should not prevent us from using it to interpret the way in which tourist practices are structured in the favelas of Rio de Janeiro. First, the evolution of the context, while it calls for the modification of Bourdieu's theory in certain aspects, does not alter the value of its concepts. On the other hand, leisure mobilities are a particular universe, in which it is not abusive to consider that there is still a legitimate culture widely recognized at both extremes of the social space. Moreover, we must keep in mind that the working classes, which would undoubtedly be the most apt to reject the legitimate model, are largely excluded from international leisure mobility. Only the middle and upper classes really participate. And among them, the sense of distinction and the deference to legitimate culture is very strong.

Achieving distinction through the practice of leisure mobility can be achieved in various ways but, today, the most important seems to be a quest for originality in the choice of practices. And this originality is achieved by choosing practices that differ from what is considered to belong to mass tourism. In this, the distinction at work in tourism is a distinction in the proper sense of the term, which is established in contrast with an "ideal type": the (mass) tourist. This is indeed of an ideal type because, in practice, the association of a given individual with the profile of the typical tourist is always situational – no one is a tourist in substance, but everyone can become one, most often from the point of view of others, by the choice of their practices, the practice thus including the destination, the mode of visit and transport, the duration of the visit, the degree of depth of their foray and even the choice of which clothes to wear. While there are individuals who, at all costs and in all circumstances, avoid absolutely anything that could lead to their classification in the tourist category, actually achieving this seems impossible, if only from the point of view of the receiving populations who, as I have already commented, rarely engage in classification operations that take up with so much subtlety the difference between the traveler and the tourist. Similarly, in practice, the meaning of the distinction from the tourist category, however sharp it may be, does not amount to a total avoidance of tourist situations, and those who claim to be travelers share, at least occasionally, a set of practices and conditions identical to tourists. This is why it is necessary to consider that, in tourism, distinction is far from operating exclusively in the domain of the choice of practices, as well as, and perhaps above all, in the discourse. It is not so much what we do that allows us to distinguish ourselves as the way in which we justify it, legitimize it and valorize it. And it is especially in the field of the discourse that the distancing of ourselves from the ideal type of the tourist takes place, mainly, as we will see, by the affirmation of a distance. Thus, it is obvious that we can be tourists in concrete terms and, for example, from the point of view of the locals, while at the same time retaining the power to legitimize ourselves in a different way, and thus to present ourselves as different from the typical tourist, whether it be with our relatives during the narrative, with our co-travelers during the practice or with an ethnologist during an interview. Many researchers who have reflected on this question have shown that "the tourist is always the other". In fact, what this point of view shows is that the tourist is not, in reality, a concrete presence that pushes us to distinguish ourselves, but a symbolic invention operated and renewed by each traveler. On the other hand, the idea that the tourist is always the other should not conceal the fact that, taken outside of individual legitimations, an individual who finds themself for their own leisure outside their home is always potentially a tourist, if only partially and conjuncturally. There would be no logical basis for the phenomenon that "the tourist is always the other" if, first of all, we were not all aware that we are tourists, or at least that we are part of a kind of practice that includes tourism, however we define it.

Moreover, not all tourist practices appear to be equally affected by issues of distinction. In this respect, visits to favelas are a special case, where the strategies used to enhance certain practices at the expense of others are particularly salient. Distinction is an essential driving force for two reasons.

The first is that the ethical dimensions of the practice are not assured. Indeed, tourism in the favelas is a matter of debate. However, in addition to the rarer criticisms that condemn all incursions of travelers into the favelas, this debate is mainly about the modes of visit and accuses first of all, rightly or wrongly, the most classic and most objectively tourist visits. I have presented tourism in the favelas as both a distinctive and ethically questionable practice. Engaging in it thus carries both potential symbolic benefits and risks. It is for this reason, firstly, that discourses and practices are strongly focused on delineating good and bad ways of visiting favelas. The considerations of ethical order, sincerely expressed by those who visit the favela, thus also form a tool available to devalue the practice of others.

The second reason has to do with the very nature of tourist experiences, which are sold and consumed as "off the beaten track" visits, in search of a reality normally inaccessible to the tourist eye. This aspect, which includes the advantage of presenting our practices as original and non-touristy, has a lot to do with the invitation to stand out. The legitimate culture of travel excludes from itself the phenomena that are too visibly consumerist, the mimetic attitudes and the idea that the valued practices can be accessible to a mass. These negative representations of good travel are exacerbated in the case of a practice that is part of a rhetoric of exploration, of crossing borders. The favela is a symbolic good, and experiencing it potentially strengthens our travel capital, as the experience increases the symbolic value of the journey made. This is what leads to a competition not for access, now open to all those who stay in Rio, but for the possibility of enhancing our own experience. In this, the distinction at work in tourist practices in the favelas is linked to the particular nature of the object, to its status (constructed by the West) of authentic otherness and, as the legitimate culture of travel would have it, inaccessible to "tourists". Leaving the strategies of distinction, it is of this authenticity that will now be discussed. Indeed, I will analyze more in depth the symbolic content of the visits, in order to show how the tourist practices in the favelas are constitutive of a construction of otherness in Western modernity. For if the favelas attract, it is not only because their visit completes the stay of tourists in search of exhaustiveness in their experience of Rio de Janeiro. Their appeal, in fact, also depends on the fact that the favelas represent a particular form of otherness that travelers want to experience. One value, that of authenticity, seems to be cross-disciplinary to the appreciation of this otherness, and it is thus of it, as a central qualifier of the relationships to otherness expressed in the tourist universe, that we shall deal with here.

5

The Authenticity of the Favela

Authenticity is an endless debate for the anthropology of tourism. Following Dean MacCannell, who introduced the question in the 1970s (MacCannell 1973), the use of authenticity to explain how tourism practices are structured has taken various forms, sometimes relating to the diversity of the objects of tourism, sometimes relating to the diversity of tourists themselves. Authenticity has thus been broken down into a multitude of forms and categories: existential (Wang 1999), hot or cold (Selwyn 1996; Cohen 2012). However, in always striving to make the domain of authenticity more precise, we sometimes lose sight of a fundamental principle: as far as culture is concerned, the idea of authenticity has no scientific basis; no culture can be authentic, because every culture is always in flux. Regarding culture, "there is no simulacrum because there is no original" (Bruner 2005, p. 5). Authenticity does not exist. Its definition, like the scarce feeling of its presence, depends on particular arrangements between individuals, groups, things and places that compose, through their encounter, the matrix of tourist practices. It is a relational and constructed value. What makes something authentic is always impossible to locate for everyone. Thus, what interests the anthropology of tourism is not authenticity as an analytical category, but the processes of attributing authenticity and classifying objects of tourism according to a rhetoric in which it plays a predominant role: that is, as an emic or indigenous category (Cravatte 2009; Cousin 2011). By what criteria is a place or group considered authentic? What does authenticity accomplish in the marketing of tourism objects? How does the rhetoric of authenticity structure the tourism of the world? These are the questions that will be the focus of this chapter.

Before coming to this analysis, it should be made clear that the authenticity in question here, that which applies to cultural entities, concerns only certain forms of tourism. Indeed, it is clear that the domain of authenticity and that of tourism in general do not strictly overlap. First, the question of authenticity goes far beyond the scope of tourism. But, above all, the tourist does not systematically question the authenticity of the objects, of the places they visit or of the groups they meet. By putting forward the concept of the "authentic fake", Brown has shown, for example, that authentic tourist feelings can arise in inauthentic places, like the famous Hiroshima Peace Memorial (Brown 1999). Authenticity is not the exclusive driving force of tourism, but as soon as tourism involves the discovery and/or encounter of cultural otherness, authenticity is never far behind. Of course, the question of authenticity is more important for certain groups and individuals than for others. But it is always there as a lever available both for the understanding of tourists and the construction of destinations. The fact remains that, in order to be allowed to speak of authenticity, the question must be directly addressed by the actors, and authenticity must be affirmed, experienced, denied and/or staged. However, the words "authentic" or "authenticity" need not appear directly and spontaneously in tourism situations. The lexical field of what constitutes authenticity may be broadened. To do this, it is necessary to clarify what we are talking about when we use the concept of authenticity and therefore, as obvious as it may seem, to underline that authenticity concerns a regime of truth. Whenever falsities and/or truths are questioned in tourist practices, we can speak of authenticity. Most often, as we will see in the case of the favelas, the true and the false – and also the superficial and the profound, the real and the staged – are present simultaneously, and it is from these dialectics, and from these dialectics only, that authenticity emerges.

One of the main aspects of authenticity is the way it is defined by opposition. In fact, attributing authenticity is an action that takes place exclusively by contrast. The value of authenticity is only relative; it is only defined within comparative statements and therefore cannot work without something that is qualified as inauthentic. Even more so, in this game of relations, it seems logical that the inauthentic must be first. We go from the feeling of inauthenticity to that of authenticity. Walter Benjamin showed it as soon as the means of the technical reproduction appeared: for there to be an original, it is necessary that there are copies. For Benjamin, it is even the copy that gives the original its aura, that is, its dimension, precisely, which is not reproducible (Benjamin 2011). All of this deserves to be recalled as this anteriority of the inauthentic over the authentic (of copies of an original) may seem counterintuitive.

However, this is a banal assertion if we want to bring together the movement at the heart of tourist mobility, from home to elsewhere, from the familiar to the exotic. Most of the time, what is considered authentic thus comes from a vague presumption based on the recognition of a true otherness. Even before tourism chooses authentic sites, it must rely on a play of contrasts, on the qualification of other elements as inauthentic. In other words, it is the inauthentic that, above all, constitutes the standard against which authenticity is measured.

Applied to its object – in this case, the favela – authenticity appears as a rhetoric unfolding on varied aspects. We will first retain three of them. The first one involves the enhancement of the favela as an authentic place, which is shown by its affirmation, by its promoters and then by the tourists themselves, as a space of otherness and authentic "Brazilianity". It is indeed the assertion of an otherness that prevails, and this is reinforced by the marginal position in which the favela was placed with respect to the formal city. In short, a rhetoric according to which the favela is perceived as authentic; preserved from a form of modernity, by the very fact of its marginality. The second aspect consists of an opposition, paradoxical as soon as the visit occurs, between the world of tourism and the world of the favela. Indeed, a large part of what is said to be "authentic" covers the field of "non-tourism". Where the presence of tourists is common, authenticity is constantly questioned. The idea of "tourist places" is often opposed to the idea, not necessarily tested, of authentic places. The promoters of the favela tours thus operate a clever staging of the exceptional character of the tourist opening of the favelas, and it is indeed because the favelas are commonly opposed to tourists that they manage to present themselves as authentic. Finally, the third type of authenticity concerns the social status of favelas as places of poverty. Indeed, the favela tours are explicitly presented, not as a negation of the stigma, but as its inversion. Poverty (but not destitution) is transformed from a negative value to a positive trait. The "culture of poverty" is celebrated, again because it is the mark of an "authentic" otherness.

5.1. The favela and the "real Brazil"

This is largely how the tourism of the favela is constructed. The favela, in fact, is presented in opposition to the formal city of Rio de Janeiro. Reading the brochures of tour operators is enough for a tourist to be absolutely convinced. In these advertisements, the favela is presented as the "real Brazil" or as the "deep Brazil". The way in which the favela is distinguished as a tourist site is particularly interesting, since the criteria evoked in the development of the idea of the favela as a place of authentic Brazil are not new aspects of what is or is not typically Brazilian. On the contrary, a large part of favela tourism consists of a reappropriation of the classic criteria that form the imaginary about Brazilian culture. This is evidenced by the insistence with which guides present the favela as the place of origin of the most popular cultural and folkloric traits in the Western imaginary.

In Romario's bar, where Favela Tour clients pass through, they serve caipirinhas that the guides call "the best in Rio", in contrast to the almost industrial production in other bars in the city. The same goes for the feijoada that is sometimes offered to tourists who have lunch in the favela. Here again, guides and tourists consistently refer to the "inauthentic" feijoadas served in tourist restaurants. Even more than in the case of the caipirinha, feijoada is presented as a homemade dish to be shared on certain occasions, and not as a dish to be ordered at a restaurant. In short, the authenticity associated with the favela rubs off and is consolidated in these small elements of Brazilian culture. They are offered to tourists as testimonies of a tradition preserved here, while lost in many other spaces in a city more and more adapted to Western standards. Regardless of the construction of authenticity, it is certain that the value of the experience of drinking a caipirinha or a feijoada is increased tenfold in the favela.

Even more important than these few typical culinary incursions is the favela's association with samba and carnival, well perceived by Bianca Freire-Medeiros (2007), which plays a fundamental role in the staging of an authentically Brazilian favela:

> The energy of Brazil is the energy of the people, the energy of the favela. It is not the rich who make the carnival. Here, in Rio, where there is the biggest carnival in the world, it is the working classes who make the show, who organize it, who play the music, who make the floats, who, how do you say? Who designs the costumes. All the samba schools in Rio are based in the favelas. Without the favelas, there is no carnival, without carnival there is no Brazil (Thiago, Chapéu Mangeira favela, August 2015).

It is clear from Thiago's monologue how the guides try their best to link the favela to certain important points in the tourist's imagination of Brazil. In this process, it is clear that the attribution of the values of authenticity is itself at stake. It is not only a question of working towards a cultural recognition of the favela, but much more of having the culture of the favela recognized as being typical of Brazil. For this, and although the tours do not actually dwell on these symbols (visits to samba schools are almost non-existent and correspond to a completely different tourist offering), it is therefore necessary for the guides to emphasize how the cultural attributes traditionally associated with Brazil (notably samba and soccer) are in essence popular elements. Since the favela embodies the popular, it is then much easier to sway the attribution of authenticity to its side.

However, the call for cultural tradition is only one aspect of the process of attributing authenticity to the favela. While the different actors of tourism in the favelas give such a great place to authenticity, it is also and above all through the implementation of a rhetoric that dissociates the favela from the tourist world. Various aspects on which the idea of the favela as authentic Brazil is developed are certainly defined in a positive and objective way, but are, however, largely dependent on an opposition with the city. Indeed, underlying this process is the presentation of the city of Rio de Janeiro as a whole as having lost this kind of authenticity. In this way, favela tourism is born and benefits from a kind of "de-exoticization" of the city, which appears to have entered progressively into neoliberal modernity and where the differences with respect to the places of origin of the tourists have been attenuated. The valorization of the favela partly works though feeding the growing absence of a tourist feeling of exoticism in the rest of Brazilian society.

5.2. Praise of the non-touristic

During a typical tourist's stay in Rio de Janeiro, nothing is as prepared as participation in a favela tour. Generally speaking, choosing a tour operator is a longer process than any other visit, and it is common for guides and their clients to exchange numerous emails prior to the excursion. Often, tourists have a set of questions to ask their host about safety, what to wear, whether to take a camera, etc. The polemics that question the ethical and moral dimensions and moral dimensions of the favela tours are also a source of profound questioning for the tourists who consider the visit. It is therefore sometimes after a long hesitation that the decision to visit a favela is taken. This is what Eric, a French tourist on vacation with his wife and two children, said:

> I was very hesitant. We're not here for very long, so I wasn't sure – especially with the kids. But at the same time, at some point, you get tired of doing only touristy stuff. The beach, Christ the Redeemer, whatever … At some point, it's good. So I said to myself: "Let's try it! Why not?" (Eric, interview, September 2015).

What is particularly interesting about this type of discourse is the way it reveals a paradox. Even though the favela tours are clearly the most "touristy" thing to take part in during an entire stay in Rio de Janeiro, the visitors manage to think of their excursion as being outside the boundaries of tourism, "off the beaten track". Because of the favela's status, the heaviness of the excursion's frame, even though it is only the presence (increasingly rare in the rest of the city) of a guide, does not harm the tourists' impression of being outside the tourist world.

5.2.1. *A perpetually renewed tourist opening*

The process of favela tourism is indeed established on the assertion of a difference of space statuses in the tourist world. Thus, the city is presented as a place where the tourist is in some way confronted only with an image that is largely intended for them. The tourist visits monuments, stays in large hotels (even in the case of youth hostels, real institutions in Rio de Janeiro) and can therefore have the sensation of evolving in a world built for their use. By contrast, the favela presents itself as a world that is diametrically opposed to it, and in which the tourist presence is conceived as an exception. The world of the favela is separated from that of the tourist by a border that the tour operators and guides offer to let them cross. In this way, the space of the favela asserts itself as more authentic, insofar as the tourist presence there is constantly rethought as an exception – unlike the rest of the city, which has a structure and a custom of reception that, for the tourist, can therefore become synonymous with inauthenticity.

When the tourist leaves the Copacabana Palace, in the heart of the famous beach, to go to the favelas, they cannot help but notice the way in which the symbolic contrasts created, consciously or not, by the promoters of the favela tours, imply a separation between the world of the tourists and what they are about to see. If the favela often manages to evoke a valued authenticity, it is largely through the staging of the crossing of a border by the tourist, from a space more or less arranged for their comfort and their pleasure towards a world in which they are this time fundamentally foreign. The interest, the pleasure or the thrill manifestly felt by the tourists in the favelas depends largely on this sentiment. Indeed, if the favela manages to be presented and experienced in an authentic way, it is because the tourist does not see, or sees in a lesser way, the pollution of their own presence. Also, the idea of the recent and exceptional opening of the favelas to tourism is constantly brought up. The Favela Tour website states that the favela "is a mystery to those who do not live there"[1]. The rhetoric of the authenticity is thus based in the staging of the mysterious, that feeling that Segalen described as a higher form of exoticism, which "hatches in the interstice of a difference, develops at the points of contact of the Diverse" (Gontard 1990, p. 21), and which for the poet was inseparable from cultural diversity: "There is no Mystery in a homogeneous world" (Segalen 2021, p. 45).

To reinforce the feeling of a visit "off the beaten track", and although they give a number of guarantees regarding the safety of tourists, the guides like to remind tourists of the indispensable nature of their presence and their accompaniment. However, the course of the excursions and the rapid dissipation of the fears that tourists can sometimes feel before arriving in the favela often lead them to think that the use of a guide was probably not, all things considered, as compulsory as they had thought:

> Mike, Irish tourist: But I feel like we could come here by ourselves. It doesn't look dangerous, actually.
>
> Leo, Favela Tour Guide: If you believe that, you can try. After all, it is a public street. You can go there. But I wouldn't recommend it. Here,

1 See: http://www.favelatour.com.br/.

> people know us, and we have this [pointing to his chest to show me that he is wearing the tour operator's T-shirt]. I'll tell you a story. A year ago, it was during the Cup, there were two Germans who came here by themselves. They started walking around, taking pictures. Young people, like the ones you see over there [obviously traffickers], started to ask them to stop taking pictures, but they didn't understand, so they kept going. And bam, one of them got shot, he died[2].
>
> Mike: Wow! Yeah, but they were probably stupid, those tourists! That doesn't mean it's impossible.
>
> Leo: If that's what you think, okay (Conversation during a tour, September 2, 2015).

If the way in which the guides put forward the importance of their presence in its protective dimension obviously responds to a strategy which aims at the conservation of their status of intermediaries, it also has the effect of reinforcing the tourists' feeling of being outsiders. Indeed, the fact of conceiving the tourist presence as rare and as requiring the protection of an intermediary contributes to the construction of the favela's authenticity. This authenticity, in this precise perspective, depends on the presentation of the favela as being opposed to the tourist's world, as not having been prepared for its reception, and even as potentially constituting a danger.

The logic of authenticity presenting the favela as a value of the non-touristic is fundamental in understanding the evolution of contemporary tourism. Valorizing the non-touristic and criticizing the touristic are two sides of the same process, which can probably be considered as inherent to the practice of leisure travel. On the one hand, this framework of values is fundamental in the evolution of tourism, at least in the West, when it is conditioned by processes of distinction, which have multiplied with particular vivacity since the democratization of leisure travel began. I have shown, in the case of tourism in the favelas, how the development of Favela Tours went hand in hand with a diversification of practices aimed at distinguishing themselves from guided tours. Nevertheless, on the other hand, it is a framework of representations, whose function is not limited to situations of comparison and distinction. This is where the concept of authenticity comes into play. Indeed, MacCannell (1976) had already pointed out, 50 years ago, that the authenticity sought by tourists could precisely be destroyed by their own presence. Condemned to always look elsewhere for an authenticity destroyed by their arrival, the tourists thus find themselves in a quest doomed to failure. The development of tourism in the favelas, in my opinion, responds precisely to this issue. This may explain the

---

2 In fact, the German tourist in question was hospitalized without serious consequences.

very attraction of tourists to the favelas, insofar as these constitute not only spaces little prepared for their reception, but above all of which one used to advocate a total avoidance. By transforming the non-touristic place into a tourist attraction and a form of heritage, the tourist economy is assured of a place of preserved authenticity. But, as tourism continues to be a threat to this authenticity, it is logical that strategies are put in place to continually restage the exceptionality of the favela's opening to tourism.

Ultimately, it is a question of transforming the favela to allow the tourist to rediscover an illusion of exploration and, to a certain extent, to escape from the shrinking of the world. This sensation of a world of which nothing or almost nothing remains to be explored or which, worse still, is besieged by tourists, is fundamental in the development of contemporary tourist practices, and in particular those which, precisely, do not think of themselves as touristy. It is, for example, at the origin of a tourist taste for risk, which is expressed in the desire to go to places that are difficult to access. But, even there, the traveler finds themself confronted with an undesired meeting with their peers. Thus, this is what Éric Boutroy indicates when he affirms that even the practice of mountaineering suffers from this concentration of travelers, and that *Himalayism*, "emblematic in this of the valorization and democratization of adventure in our society, has known a relative trivialization for some decades" (Boutroy 2006). It is perhaps also from this point of view that we can understand the increase in tourist interest towards urban margins, which come to embody the unknown, the non-touristic and the preserved. The romanticism, closely linked to the act of travel, is continually fed by travel literature, by fictions that feature courageous explorers, impeccable adventurers. However, the tourist who sets out to encounter the world often only finds the traces of the passage of their peers and the changes that follow, namely, the standardization and professionalization of hospitality. It is in this perspective that the tourist value of the favela is constructed, that is, in the measure in which it manages to embody a territory to be discovered, whose exploitation requires courage, is accompanied by uncertainties, and on which – it is a rare advantage – the narratives of the professional travelers have not already multiplied. In fact, as I have already mentioned, visiting the favela is accompanied by a certain prestige, which is perceptible at the moment of talking or writing about it. In a world perceived as increasingly uniform, urban interstices and what sociology has called subcultures constitute appreciable entities for tourists in search of otherness. Let us consider the calculation proposed by Gérard Leclerc concerning the shrinking of the world:

> It took Magellan's sailors exactly three years (September 1519–September 1522) to complete the first circumnavigation. The Earth is, for the jet set, 500 times smaller than for Magellan's sailors (more precisely, these times are in the ratio of 365 days × 3 = 1,095 days to 2 days, or 548/1). Let the reader weigh up and contemplate

> these figures. From the point of view of the most technologically "advanced" humanity, planet Earth is 500 times smaller than in the Renaissance; 40 times smaller than in the 19th century (Leclerc 2015, p. 141).

While it is thus clear that, with the (relative) exception of the deep wilderness of the poles and the great equatorial forests, the shrinking of the world is at an advanced stage; is it not then the idea of a world still to be discovered that the urban margins seem to embody? Is it not in the interstices of the urban civilization, rather than at the borders of the world, that we manage to construct an otherness with a powerful force of attraction?

5.2.2. *The denial of merchants in the favelas*

One of the most touristic aspects when visiting the favela is obviously its commercial character: tourists pay to have access to it. When authentic otherness is the stake of the practice, its protagonists face a paradox. Indeed, tourism trades in culture, while dealing with an imperative to dissimulate this trade. Tourism thus always develops in an ambivalence between enchantment and the marketing of the world. Bertrand Réau has highlighted this condition of tourism work and of what he calls the logics of re-enchantment of the world, coupled with a denial of the commercial character of the experience. Involved are the multiple strategies employed by tourism intermediaries to neutralize the question of money:

> One of the conditions for the proper functioning of the tourism market lies in the denial of its immediate commercial character. One cannot then accept as self-evident the idea that tourism would call into question the norms and values of daily life, with the sole social function of recuperating and restoring labor power. Placed in the space of cultural practices whose specificity is to produce an enchanted relationship to the world within a market relationship, the suspension of the ordinary world in the tourist situation constitutes on the other hand a starting point for reflection (Réau and Poupeau 2007).

Anne Doquet provides a compelling example of this, showing how guides in Mali conceal the remuneration of locals who participate in the tour, leading them to believe that this participation is completely free, just as they themselves conceal the commissions they receive from the sale of objects (Doquet 2009). As for the content of the tour (apart from optional purchases), tourists usually pay a price in advance that includes any payments to locals, which cannot exist in the eyes of all, at the risk of destroying any sense of authenticity. Concealing money is common in tourism, especially when it comes to meeting an otherness for which a rhetoric of authenticity dominates.

The relationship maintained with the market economy in the favela tours has its particularities, however, and the denial of the economic is not done in quite the same way. This is partly due to the way in which the tourist economy has been structured in the favela. I have shown that the guides had a great deal of control over the development of tourism in the favela and that any other type of participation in the reception of tourists was subject to their intermediation. The direct remuneration of other individuals in the favela is rare but does exist, especially in the form of renting terraces, which allows the group to have a view of the favela. In this case, it is not uncommon for the rental to be evaded and for the guides to present the owner (if they happen to be present at the time of the tour) as someone who kindly opens their doors to tourists. However, the guides do not pay the capoeira dancers, restaurant owners, other locals who open their doors, artisans, vendors, etc. The deal that the guides make with some of the locals is limited to getting the groups to circulate near them so that they can deploy their own commercial strategies. But, as proof that the guides are aware of the need to orchestrate a denial of the commercial nature of the favela's discovery (or, at least, to limit its presence), these partnerships are selective and limited. Marcelo, founder of Favela Tour, has had to move cautiously in this direction. He recounted:

> I make sure there is only one commercial point in the tour. In Rocinha, we go to the artisanal market, it is good. There are people who work there, it helps them, it's very good. But, at some point, in Vila Canoasthere, another small market was set up, and there it was no longer possible. It's too much (Marcelo).

In cases where it is difficult to avoid them, it is not uncommon for guides to reassure tourists by anticipating solicitations and reminding them of their right not to buy. As a general rule, the issue of money is addressed rather explicitly and only seems to create discomfort in the case of payments at the end of the tour, when the guide has to break with the friendly image they have built up during the tour, reminding people that they are the provider of a service. Before and during the tour, the issue seems to be resolved by the same strategy of defusing ethical concerns. Indeed, the promoters of favela tours cannot simply evade the question of money, since the idea of trading in an otherness such as that of the favela poses an ethical question in itself. And, in a sense, they manage to deal with the ethical problem and deny the commercial character simultaneously. First, the use of the funds generated by tourism is always on the agenda: no tour operator or independent guide can carry out its activity without establishing a partnership with an association, whatever its purpose and even though it is itself the organizer. The causes defended can be varied, as long as they serve the second and main condition: that of working for the symbolic valorization of the favela and its inhabitants. Indeed, what the favela tour promises is to take the favela out of the stigma of which it is the victim. The empowerment logics put forward by the guides have an essential weight to counter possible ethical concerns. The observation of the excursions shows that this weight is more important than that of the strict economic. The tourists, moreover, are not fooled by the weight of their financial contribution, whereas their symbolic participation is considered important. In short, it is by circumstance that the denial of the commercial character of the favela tours does not operate a complete dissimulation of the financial aspects. But this denial is still visible in that the considered effect of the excursions, their profits, is to be sought on the symbolic side. The guides say to the tourists: "You do not enrich the entrepreneurs, you do not even really enrich the favela, you improve the image of the favela".

5.3. The authenticity of poverty

The idea of poverty is inherent to the image of the favela. Contrary to other preconceived ideas that the guides think are shared a priori by tourists (crime and violence, in particular), it is not a question here of totally deconstructing the association between the favela and poverty. Indeed, this is part of the tourist appeal of the favela. This is what researchers point out when they place favela tours in the category of poverty tourism, thus making poverty the central element of these practices. Still others describe favela tours as poorism, thus giving the relationship of tourists to the object of their visit a discriminatory inflection. The idea of poorism is systematically accompanied by that of voyeurism and lies in the idea of the transformation of a place of life for the poor into a gloomy visual consumption. It is here in a different perspective that I will present the place that poverty occupies in the tourist practices of the favelas. Without denying that poverty is a determining element of the visits, it is a question of subordinating the question to that of reality, contained in the category reality tourism, about which I have already stated my preference. Indeed, it seems that if poverty has an important place in favela tours, it is mainly as a quality that conditions the authenticity of the place and its inhabitants. It is only in this context that poverty appears as an element that can be valued by tourist practices. However, there is an asymmetrical relationship that conditions the tourist discourse on poverty and constitutes the mark of a fantasy about poverty.

5.3.1. *The spontaneity of the favelas*

The favela is initially perceived as a poor area, and many tourists have an image of a miserable slum. However, it should be pointed out that the favelas visited by tourists are not the poorest, neither among the favelas nor among the disadvantaged areas of the city. Favela poverty is first presented by the guides as a cliché to be deconstructed. The presence in favelas such as Rocinha of a middle class, high definition televisions in the living rooms, four-story buildings, well-dressed inhabitants on their way to work are all images on which the guides rely to show the discrepancies between the stereotype and reality. Poverty, however, is not a minor theme of the visit, and the guide's discourse consists of fighting the association of the whole favela with poverty, not in denying any form of poverty. Moreover, it is the term "destitution" that is explicitly proscribed by the guides:

> You probably think you're going to see destitution. But you will be disappointed. People from the favelas are not rich, but they don't live in destitution, as you'll see, and they are happy to live here (Zezinho, before a tour in Rocinha, August 2015).

Once the destitution is removed, there remains a poverty that is no longer abominable and that becomes not only morally demonstrable, but also valorizable, which is what a large part of the visit consists of, which is done on the model of a symbolic inversion. This inversion consists of keeping the same elements (poverty, marginality and, in some cases, criminality), but transforming them into valued elements, because they represent an otherness. In order to develop, the valorization of poverty is rooted in the evocation of different values: spontaneity, creativity, freedom, etc.

The meeting with the inhabitants being almost inexistent, the discourse of the guides applies, for the valorization of the favela and poverty, on aspects which are primarily visual. In this case, the particular architecture of the favela holds an important role. This can be seen in the following account of an excursion:

> [After an hour of touring, we find ourselves in an alleyway in Rocinha, which our group of six is walking through behind Erik, our guide for the day. Arriving at a strategically chosen point, Erik turns to the group and begins to explain a few elements of the setting].
>
> Erik: Here, there are a few things I'd like to show you. First, look above your heads. What do you see?
>
> [The tourists look up and search for a few moments, until one of them, a German tourist in his fifties, speaks up].
>
> Tourist No. 1: Cables?
>
> Erik: Yes, that's right. I'll talk about that in a moment, but what else?

[Tourists continue to search, to no avail].

Erik: Here, in the alleys, you'll notice that you can't see the sky. You will experience this when we continue our way down the favela. The reason for this is that there are no rules for building in the favela. There is no minimum distance between the houses. Look again, you don't see that anywhere else, certainly not in your house.

[The majority of the group continues to search the sky in vain, sometimes taking a few steps backwards. Others take pictures. A Danish tourist speaks up].

Tourist No. 2: But who builds here?

Erik: Usually the people who live there. It goes like this: whoever owns the second floor can decide to sell their roof. The one who buys can then build a floor on top of it, and then sell his roof, and so on, up to four floors which is the regulation imposed by the government. But this is not respected and here in Rocinha, and we have buildings with seven or eight floors.

Tourist No. 2: Wow!

Erik: It's usually the residents themselves who do the work, or get other residents to do it. You see, a lot of the men here work in construction, so they have the know-how to build houses. Look behind you, in front of me.

[Erik points to a stone staircase with mosaics on it that leads directly to the second floor of a house].

Erik: Here is a man I know who made his stairs. See how beautiful they are?

[While some take pictures of the stairs, Erik continues his explanation].

Erik: I saw him make this staircase. In general, people take advantage of the weekend to do work on their homes, they try to make their homes more beautiful.

[Erik gives the group a few moments to examine the buildings, then resumes].

Erik: Earlier, I told you that the favela was entirely autonomous. Rocinha is entirely built by the inhabitants, and they are the ones who bring in the main services. Look at the electric cables that our friend was talking about and look at the pylon behind me.

[The tourists, some of whom have already had time to take a good look at the power pole Erik is standing in front of, are waiting intently for his explanation].

Tourist No. 1 [camera in hand]: It's impressive.

Erik: Yes, the electricity here is diverted. These are cables that are pulled from the city. When you build a house in the favela, you have to find the nearest pole and pull a cable to your house. The same goes for the pipes. You can see that I am on a plate.

[Erik moves slightly and lifts the plate he was standing on].

Erik: Here you can see the favela's sewage system. Everyone has to bring their pipe to these relays.

[As Erik and the group continue the tour en route to another point of interest selected by the guide, I chat with Adrien, a French tourist who showed curiosity about my research when Erik briefly introduced it at the beginning of the tour. Adrien, passionate about photography, does not miss an opportunity to take a shot of this unique architecture and various signs of the favela's particular organization].

Me: I see you take a lot of pictures.

Adrien: Yes, I am a photographer. I will make an album of my trip.

Me: You're going to be able to take some interesting photos here!

Adrien: It's clear. It's great here. The architecture is fascinating. It's completely spontaneous, a kind of anarchy but organized. I didn't expect this at all.

Me: Why? Were you expecting something specific?

Adrien: No, not really, but I imagined something more precarious.

Me: Wooden shacks and tin roofs type?

> Adrien: Yeah, a little bit. But I've already seen pictures. But here, to see it from the inside. Just now, at the top, I thought my eyes were playing tricks on me. It gives a crazy landscape. It's really beautiful, actually. I mean, even from the outside, it's beautiful. But now, it gives a different point of view.
>
> [During our discussion, Erik stopped at our level to listen].
>
> Erik: I'm glad you like it. I don't know if I told you, but I've had groups of architecture students come and observe a few times.
>
> Me: Oh yeah? Brazilians?
>
> Erik: Oh no, never as you'd suspect. Europeans. Americans, too. They are interested in the architecture of the favela, which is spontaneous and adapts well to the particular environment, the *morro*, and all the solutions that had to be found, and without the help of architects, just builders who, with the know-how and the ownership of the land, managed to do that (Conversation recorded during a tour, Rocinha, July 2016).

It is certain that the valorization of the spontaneity of the architecture, at work in the tourist excursions, is a phenomenon that breaks with a perception that has been dominant for a long time. If we take into account, in fact, the health report from the first half of the 20th century, and therefore the initial denunciation of the favelas as "miasmas of the city" (Valladares 2006), we cannot help but notice how these tourist perceptions constitute a novelty. The adjectives Adrien used to describe what he liked about the landscape are a good representation of tourist valorization. Indeed, the idea of "organized anarchy" and the central concept of "spontaneity" are not insignificant and correspond to the discourse of the guides on the favela, although at this stage of the visit, they cannot be considered as a conscious reiteration of tourist slogans. What is emphasized, especially through the architecture, is the idea of an autonomous community that shows extraordinary inventiveness in the face of adversity. The obvious pride that Erik shows when he talks about the architecture students' visit is something I have observed in many guides who have had the opportunity to lead such groups in a favela. All of them, in fact, congratulate themselves on behalf of the community for the mystery of the favela's architectural development. They sometimes add:

> They come to try to understand how it works, how we manage to make these buildings stand in such difficult terrain. Even the architects in town don't understand (Marco, local guide, Cantagalo).

It is largely in this spontaneity that the aesthetic value of the favela is rooted. Opposed to the straight lines and the squared districts of the *asfalto*, the favela seems of an infinite charm in its curves, its irregular entanglements and its development as chaotic as harmonious.

5.3.2. *A culture of poverty*

More than an uncontrolled development, what the guides affirm is the idea that behind the spontaneity of the construction, there is not a state of fact, but a desire, even a real culture. Consider, for example, the story that guides often tell about the history of Cidade de Deus. This is how Obi tells it:

> There is a favela that you probably know called Cidade de Deus: City of God. I don't want to tell you about the movie, but just the story. Cidade de Deus is not a favela – at least, it wasn't originally a favela. In the 1950s, the government destroyed several favelas and relocated the people to a new neighborhood of small houses. There were streets, each house had a garden – like an American suburb, a little one – but of course, not luxurious at all. But little by little, people continued to do what they did before in the favela. One added a floor, the other did the same. Other people came and built next door. Then they started to pull cables and, little by little, it became a favela! You see, when you are a favelado, it's in your blood. Here, they often say: you can leave the favela, but the favela never leaves you (Obi, during a visit to Rocinha).

This story, common on the favela tours, if only because it allows for a reference to the famous film, is heavy with undertones. Clearly, one of the objectives is to highlight a critique of the state, since the example of Cidade de Deus shows at once two typical government postures: its willingness to "eradicate the favelas" and then its lack of investment in the renovation and care of the city's poorest social classes. But, and especially through the way the event is told by the guides, they also imply another idea, this time directly related to the favela population. What they imply in this type of story is the existence of a real favela culture. Thus, if the Cidade de Deus has become a favela, it is not only because of the abandonment of the state, but also because of a tendency of the inhabitants whose culture pushes them to transform their habitat into a favela. Of course, it is in a positive dimension that the guides understand the responsibility of the inhabitants during "favelization" of the Cidade de Deus. In fact, it is only after having already worked for the cultural recognition of the favela and its occupation, not as an undergone social condition but as a valued way of life, that the guides generally evoke this episode in the history of the favelas in Rio de Janeiro. Moreover, it is quite common to find a completely different interpretation of "favelization" in public opinion, and there are many in Brazil, for whom favela culture is not a creative impulse: on the contrary, it represents for them a shortcoming that makes the problem of the favela insolvable.

Here, we are faced with a set of representations which, in my opinion, bring about the concept of the "culture of poverty" in all its complexity and inherent contradictions. Developed in the 1950s, notably through the work of Oscar Lewis, the concept of the culture of poverty covers a set of studies devoted to the working classes, thus recognizing the existence of a language, a culture, representations and principles specific to groups in a situation of poverty (Lewis 1959, 1969). Here is how Lewis, in the postface to *La Vida*, defines the concept of the culture of poverty:

> As an anthropologist, I have tried to understand poverty as a culture or, more precisely, as a subculture, with its own structure and justifications, as a way of life transmitted from generation to generation. This conception draws attention to the fact that the culture of poverty in modern states is not only a matter of economic destitution, disorganization, or lack of something. It is also a positive element that has its good points without which the poor could hardly hold out (Lewis 1969, p. 801).

The idea, mainly found in English speaking countries, has since been gradually abandoned and has rapidly become difficult to use. Indeed, Lewis' work, in addition to a specific cultural recognition of the working classes, also intends to explain the culture of poverty as a mechanism that contributes to the reproduction of inequalities, and that tends to lock the poor into their condition. The idea, while not necessarily problematic in itself, has suffered from its political recovery, notably by certain American conservatives who have positioned in it a scientific argument to explain the responsibility of Black people in their own marginalization from society, and more concretely to do away with the social assistance on which they would have become completely dependent (Duvoux 2010; Small, Harding & Lamont 2010). As a result of this political usage of the concept, which is far removed from the principles of sociologists and anthropologists who tried to delimit a culture of poverty, the concept has almost disappeared from the social sciences[3]. However, it is worth using in order to understand tourism practices in relation to poverty although it is no longer a question of considering a culture of poverty as an analytical category for anthropology, but rather as an indigenous category, that is, as something that makes sense in a certain context of interpretation of the world, in this case tourism.

---

3 Some sociologists, mainly American, have never totally abandoned the notion. William J. Wilson, for example, has been asserting the need for it since the late 1980s (Wilson 1987).

Indeed, as I have shown through several examples, tourism in the favelas is largely articulated, from a Western point of view, around a cultural recognition refused by the rest of the Brazilian society, and more specifically around a valorization of the strategies of adaptation when faced with exclusion and poverty, as well as the way of life and the nature of the social links in this condition. This is where we can find the most relevant way to link the different tourist phenomena affecting the stigmatized popular neighborhoods around the world. In other words, the recognition of a culture that is not, or at least not only, specifically Brazilian, Indian or South African, but of an (urban) culture of poverty being the element that best allows us to link them. In this, the culture of poverty is well perceived[4] as having a universal dimension. At many points in the tours, there are signs of this general interest in the condition of the inhabitants and the recognition of a culture that is specific to them, and which allows them to be perceived not as being specifically inhabitants of this or that favela (let us remember that each favela is valid for all favelas, in a tourism context), but as a social class that develops in conditions that are foreign to tourists and which, in many aspects, can be the object of a valorization.

Of course – and this is a point of insisting that the concept of the culture of poverty is addressed here exclusively as an indigenous category – the cultural recognition and valorization I am attempting to analyze is largely the product of a fantasy about poverty. If we could safely use the vocabulary, now probably a bit outdated, that Richard Hoggart uses in his work on English popular culture in the 1950s, we would present the tourist valorization of the favela as the result of a bourgeois fantasy about the proletariat, and of what the British anthropologist calls a "populist" attitude:

> From compassion ("The common people would be perfect if ...") to glorification ("They are always perfect ..."), the range of populist sentiments covers a whole series of novelistic myths, beginning with the Lady of Bath: the popular classes are fundamentally "healthy" – such is the initial literary theme – much healthier, on close inspection, than the other classes; the common people are like diamonds, rough and unpolished perhaps, but isn't that better? They are a little rough, but "worth their weight in gold"; they are neither intellectuals nor

---

4 This is, once again, the culture of poverty as an indigenous category. As far as its scientific use is concerned, the most eminent researchers who have worked on the concept have always been wary of the idea of such a generalization and have often been content with a well-defined regional, sometimes national, context. For example, Oscar Lewis writes, "A distinction must be made between countries where it represents a relatively small fraction of the population and those where it affects a very large part of the population" (Lewis 1969, p. 809).

> refined, but they at least have their feet on the ground; they know how to laugh without constraint, kind hearted and hearts of gold (Hoggart 1957).

The idealization of the favela's cultural and social identity is not far removed from Richard Hoggart's literary analysis of bourgeois representations of the working class in 1950s England. The idea of a kind of poor man's wealth is indeed present at various points in the symbolic consideration of the excursions. Also, the impressions of some tourists collected a few hours after the tour are in line with what Hoggart describes:

> I was fascinated by these people. When I go to a country, it's to meet the people. I'm not interested in Americanized, middle-class people. Frankly, here, [we are in a bar in Copacabana, a few blocks from the beach] look, it's like in Paris: the waiters are pouting, the people seem to be in a hurry, and everything. Earlier, in the favela, it was something else. People were smiling, they were in the street drinking, quiet. They have nothing, but they look a thousand times happier than us. It makes you think, that's for sure (Ornella, French tourist, the day after a tour in Rocinha, August 2016).

The discourse of the guides, too, sometimes goes in this direction. Thus, when Zezinho evokes the mentality of the members of this community, he emphasizes a posture that could certainly appear as that of a resignation regarding their condition, but which he intends to set up like a philosophy:

> People are not unhappy here. We are content with little. I wouldn't want to live anywhere else in the world. I have everything I need here. I don't need to become rich. Here, life is not expensive. And there are always people helping each other, etc. (Zezinho, reflecting on a trip to Rocinha).

Several important ideas, already mentioned, are concentrated in this discourse. First, there is the idea of a satisfaction linked to the way of life of the favela's inhabitants. Contrary to the generally accepted idea, the inhabitants of the favela do not desire to leave. Second, and this is obviously related, this type of discourse implies the existence of a causal link between a modest way of life and a particular strength of the social bond. Here, what is thus co-constructed by the tourists and the guides is an opposition between the favela and the tourists' world, which articulates around a setting in contrast of the nature of the social bonds in both spaces. It is almost the old Durkheimian opposition between organic solidarity and mechanical solidarity (Durkheim 1930) that is materialized in favela tourism. Indeed, the latter takes, in the tourist context, the shape of a space of strong socialization, of a permanent solidarity and of a daily life experienced in a community. In this perspective, the preferred use of the term "community" at the expense of "favela" is significant and allows the guides to emphasize the powerful links that unite its members. Very inclined to critically distance themselves from their society of origin, tourists then find it easy to imagine the inhabitants of the favela as strangers to the problems of individualization that constitute their daily lives:

> Yes, I think that, after all, it can be good to live in a favela. Of course, I'm not talking about places where there is violence and trafficking, nor living in poverty. But here, life looks peaceful and people don't look miserable. And it was explained to us that people here do not let themselves starve, they help each other a lot. I think that's what's missing here – in Europe, I mean. I think we are very lonely, in fact. We work, we take the subway, we watch TV and we are all alone. We don't help people enough and we just don't talk to each other enough (Angel, Spanish tourist, during a tour in Santa Marta, August 2016).

The idea of a mechanical sociability of the favela is thus often evoked on the tours. It is indeed opposed to modern capitalist societies, in which solidarity is regulated by structures that are mainly state-owned. But, of course, the mechanical/organic opposition that leads to the valorization of the favela is not only, or even mainly, about solidarity, but more generally about what this absence of mechanization (i.e. concretely, support from a public political body) implies about the daily lives of the inhabitants who are therefore more interconnected, forced to rely on the whole community. What this opposition also implies is the existence of a strong sense of belonging, which can be seen as a powerful attribute of the culture of poverty. By emphasizing the social cohesion among *favelados*, the guides play on the sense of individualization felt by many tourists, and whose uneasiness is often expressed in relation to the desire to distance themselves from everyday life through the practice of travel. Within the framework of the tourist excursions, the supposed power of the interconnection of the favela's inhabitants is never commented on other than through a flagrant idealization. Not once is the idea introduced that this organic sociability can sometimes be a source of discomfort for the members of the community and an integral part of what some of them may sometimes want to escape through a change of condition.

In addition to these few main principles of the imagined way of life of the favela inhabitants, the guides also give an important place to the recognition of certain values typical not only of the favela, but also, in my opinion, of poverty as a social condition. As a result, the favela tours develop around the recognition of a culture of poverty that tends to valorize the poor. If we try to go back, in a general way, to the bourgeois prejudices concerning poverty, as described by Hoggart, we realize that tourism in the favelas indeed displays a positive perception of the poor, of their potential for innovation and of their way of life. This is related to what Lewis wrote:

> Throughout written history, in literature, in proverbs and popular sayings, one encounters two contradictory evaluations of the nature of the poor. Some people consider the poor to be good, virtuous, just, serene, independent, honest, kind and happy. Others see them as wicked, evil, violent, sordid and criminal. These conflicting and confusing views are also reflected in the internal conflict within the war on poverty. Some emphasize the great potential of the poor to help themselves, to govern themselves, and to organize in communities, while others stress the destructive and often irreversible effects of poverty on the individual, insisting at the same time on the need to leave control in the hands of the arguably more mentally healthy bourgeoisie (Lewis 1969, pp. 800–801).

If we agree with Lewis' statement, that is, if we consider that two opposite visions of poverty coexist, then we can situate the favela tours on the side of the first type of representation. Indeed, as we have seen, the actors of the favela tours co-produce a positive representation of the favela inhabitants, even going so far as to envy certain aspects of their way of life to tourists. This is a fantasy, of course, and we cannot help but view with skepticism the claims of tourists who say they are tempted to adhere more permanently to such a condition. Moreover, when tourists become temporary inhabitants of the favela, it is obviously under certain conditions, which are not without giving them a high place in the social hierarchy of the neighborhood, assuming that they are one day really included in this hierarchy. Long-term travelers from Europe or the United States, students, guides and entrepreneurs, who generally occupy some of the finest houses in the favela, adhere to a way of life and a culture, but without really sharing the living conditions of their new neighbors. In any case, it is certain that their attitude, like that of the tourists, is characterized by a positive view of the poor and, fundamentally, by a real political and emotional interest in the issue of poverty.

The hypothesis of a tourist identification with a culture that is one of poverty is therefore fundamental, not only for the understanding of the tourist practices that I analyze in this work, but also in the perspective of a rapprochement between what can be observed in the favelas of Rio de Janeiro and what is happening in other large cities of the world. This logically assumes that the foundations of tourism in these spaces are not found, in any case decisively, in their specific national conjunctures, but in what links them: the modern urban poor condition.

The culture of poverty thus seems to be the main object of tourism in the favelas. Their valorization, in fact, depends largely on the positive values associated with poverty: sincerity, solidarity, honesty, spontaneity, independence, resourcefulness, originality, creativity, etc. These values hold a fundamental place in the process of transforming favelas into tourist sites. The urban poor condition has in itself a tourist potential insofar as, from the point of view of tourists who do not belong, or no longer belong in any case, to the same social classes, it represents a preserved otherness. Indeed, the moral importance of cultural diversity today leads us to the recognition of an otherness, present even within large metropolises, which is now increasingly perceived as a culture, and not only as a subculture or, even worse, as a part of the population alienated and deprived of all cultural particularities. The enchantment of the world through the celebration of cultural diversity makes the culture of the poor a prime tourist object.

In the end, the logics of tourist valorization applied jointly by the professionals (to promote) and the tourists (to valorize the experience) are developed by inserting this culture of poverty into the economy of authenticity. Indeed, the values of authenticity largely correspond to what is valued in the type of spaces that the favelas of Rio de Janeiro represent. As a break with the everyday and the domestic, the journey promotes the construction of favela representations, and more generally of poverty, as opposed to the negative aspects of modernity. Faced with the alienating dimensions of modernity, the tourist is prone to think of poverty and exclusion as sources of an authenticity that they have lost in their daily life. Richard Hoggart's depiction of the working-class prejudices of the Marxist bourgeoisie in 1950s England demonstrates the common association between poverty and authenticity:

> For the sake of the feeling of pity that he [the Marxist of bourgeois origin] feels for the "alienated" worker, he is led to attribute all his faults to the oppressive system to which the working class is subjected. He often admires what would remain of the good uncivilized among workers. He keeps the nostalgia of these artistic manifestations "superior to all the others" that were the peasant art or the "authentically" popular worker art and he manifests a very particular enthusiasm for the vestiges of this culture that he imagines to discover in contemporary life. He deplores and admires at the same time the traits of Jude the Obscure that he lends to the workers. In fact, mixing pity and condescension, he generally succeeds in giving and giving himself a totally distorted view of the working class (Hoggart 1957, p. 41).

I have constantly shown, when talking about authenticity, that it is a vague notion whose attributes are defined above all by opposition, that is, by starting from an assertion of inauthenticity that pushes us to look elsewhere for what we consider lost, even though, without a doubt, the object of our nostalgia has never existed. In the context of tourism, the notion is central, insofar as without authenticity, the whole trip and the prestige that we can potentially derive from it are jeopardized. The things seen in Brazil, in fact, have a real value only if "they are authentically Brazilian", that is, if they cannot be found elsewhere. This is how authenticity works when we try to associate it, in a tourist situation, specifically and concretely with something. But this is not always the case. Often, tourists do not seek a singular identity, but simply an otherness. What is then authentic is what is authentically different – authentically exotic, we might say. Poverty tourism, when it is most effective, manages to capture both dimensions of authenticity. I have just shown how the culture of poverty manages to embody an otherness preserved by the very conditions of its exclusion. But this cultural preservation also works to associate the favela with something typically Brazilian. Not only does tourism recognize a culture of poverty, but also it conceives it, at least for the time of the visit, as more authentic through the argument, often used by guides, of the importance of popular culture.

In this section, I have so far tried to clarify the place held by favelas in the world of tourism. Obviously, favela tours are part of a rhetoric of authenticity, which leads in large part to the permanent staging of the favelas' openness to tourists. It is because they manage to think of themselves as tourist exceptions, or even as a departure from the beaten track, that favela tours have undoubtedly become so firmly established in Rio de Janeiro's tourist offer. But the idea of authenticity, fundamental as it is, is not enough to explain the co-production of a tourist favela. Even more, it seems that the preferred use of the concept of authenticity conceals definitively a tourist posture that is not so far from exoticism. And if the favela succeeds in imposing itself as an exotic destination, it is largely because its inhabitants – and we must understand here not only the *favelados*, but also and especially the urban poor – now embody an otherness that fascinates. It is therefore to the discovery of this way of life that the tourists who stay in Rio (but also in other big cities of the South) launch themselves. This is an otherness whose representations I have shown to what extent they can depend on a bourgeois, even primitivist idealization, the poor, to use Hoggart's formula, embodying in some way "what remains of the good uncivilized" (Hoggart, ibid.). Unlike the large cities of the North, where the practice remains more than marginal, the metropolises of the South, as they develop and enter the neoliberal dance, seem to lose their capacity to embody the exoticism necessary for tourist pleasure. In this context, the urban interstices, those of poverty and exclusion, constitute the last bastions of a lost cultural integrity. The favela embodies a truly exotic Brazil in the face of "beaches full of tourists" where there are "only white people".

5.4. Tourism, slum and poverty

Studies have shown how, in other national contexts, urban spaces of poverty can be placed in a tourism context. Contemporary tourist curiosity about poverty is not in itself a new phenomenon, although it has only recently reached the dimensions of mass tourism (at least for Rio de Janeiro and Cape Town). The best documented origins date back to the second half of the 19th century, in Victorian London. Visits to East London's slums and working-class areas by middle- and upper-class individuals do have something to do with the practices observed in this research. In 1884, an article in the *New York Times* was devoted to the new trend of "going to people of whom they have heard, but of whom they are as ignorant as if they were inhabitants of a foreign country"[5]. Already, these transgressions of urban boundaries by the more affluent social classes are a complex phenomenon that covers various aspirations. On the one hand, part of the phenomenon manifests a willingness on the part of the wealthiest to "slum it up" and an obvious curiosity for the mores of the poor, but the phenomenon is equally indissociable from a multitude of philanthropic ventures and a real desire, if not to help the poor directly, to learn about the problem of urban poverty in order to find ways to remedy it. We cannot help but notice the continuity that characterizes the double facet of these relationships maintained by the upper classes with the urban poor. A source of fantasy and at times valorized, disadvantaged neighborhoods are also the receptacle of more or less paternalistic desires to help fellow citizens, and thus, the site of "unasked-for volunteerism" (Koven 2006).

As many things in tourism, slumming is also a practice of British origin and has been documented in various times and spaces, the most illustrious and best commented on being New York at about the same time, notably in the famous immigrant neighborhood of Five Points or in the Bowery, then at the beginning of the 20th century in Harlem, where people went to listen to jazz and take advantage of the easy access to drugs (Novy 2011), and also in the heart of World War II, in the Warsaw Ghetto (Ringelblum 1995)[6]. In the way these practices manifest a tourism of the "culture of poverty" as well as in their very nature of leisure mobility (as opposed to a forced or, at least initially, non-negotiated reception), it is certain that we can see a forerunner to the practices currently studied by many researchers elsewhere than in Rio de Janeiro: in Mexico City (Dürr 2012), in Johannesburg, in

5 See: https://www.nytimes.com/1884/09/14/archives/slumming-in-this-town-a-fashionable-london-mania-reaches-newyork.html.

6 In the case of Warsaw and because of the context, the connection with slumming is somewhat dangerous, insofar as it is less a culture of poverty than a willingness to see war that seems central. More often cited as an example of thanatourism or dark tourism (Stone and Sharpley 2008), the case highlights a certain morbidity that is irrelevant to understanding the motivations of modern slummers, who work to enhance the spaces they visit.

Cape Town (Rolfes et al. 2009), Mumbai (Dyson 2012; Meschkank 2011), Nairobi (Chege and Mwisukha 2013), Caracas, Cairo (Mekawy 2012), New Delhi, etc. In general, it is all of the peri-urban areas of the South that progressively attract the interest of tourists from the North, finding there the last bastions of a radical otherness. To forge this otherness and reinforce the exoticism, it is certain that poverty (a condition that brings together, although not exactly on the same level, all of these spaces) plays a central role and that there is indeed a recognition of a culture of poverty. What the Victorian slumming of the late 19th century and the resurgence of the phenomenon in the 1990s have in common, almost simultaneously in Brazil and South Africa, it is a disposition of members of wealthy classes to recognize not only the existence of a culture of poverty of which they know nothing about, but to value it, even to overvalue it, at least during the period of a tour.

However, despite their filiation, the two phenomena present notable differences. First, as Fabien Frenzel and Ko Koens (2012) aptly note, it is above all the perceptions of the social functions of tourism that have changed. Even though Victorian slumming already had a dimension that could anachronistically be described as humanitarian, tourism is now perceived more and more broadly in a solidarity perspective. On the contrary, the phenomenon I comment on in this book is incomparable in terms of its scale. In 2014, an estimated 1 million tourists participated in tours of spaces commonly considered to be slums (Frenzel 2016). Far from the few bourgeois coming to experience slumming in working-class neighborhoods, the practice of *slumming* has now reached a level of massification and globalization. Finally, and this is a fundamental difference, it is indeed in the context of international tourism that the massification has taken place. Whereas in the London context of the 19th century, as in Harlem a little later, it was almost exclusively a matter of crossing borders in their own city, the phenomenon is the reverse today. The contemporary cases cited above, which researchers have grouped together under the label of slum tourism, show an astonishing evolution, insofar as it is mostly foreign tourists who come to visit the segregated neighborhoods. While it is inconceivable that international travelers have not also been interested in the thrill of slumming in both London and New York, their participation could only have come after locals from the same classes had, in a sense, paved the way before their arrival. In contemporary cases, and particularly in Rio de Janeiro, the opposite seems to be the case.

The valorization of the working classes, through the valorization of their spaces, is in fact the fruit of the representations and practices of international tourists and finds few possibilities of rooting itself in local representations. Moreover, the development of tourism in the favelas of Rio de Janeiro, in the early days, pitted tourists against locals from the same social classes. For many local members of the middle and upper classes, as I have had the opportunity to meet many of them, it seems inconceivable that tourists see in the favelas anything other than the possibility of a thrill, and it is very rare that a real cultural dimension is associated with these practices that they consider strange, not to say naive. On the tourist side, the bourgeois carioca is identified as an enemy and held responsible for the stigmatization suffered by the favelas, and which prevents the recognition of their culture. A significant part of the guides' work consists precisely of placing the tourists in this opposition, even though it is doubtful that most of them display the same disposition to the cultural recognition of the most popular urban classes of their own city, nor even, in the end, to really ask questions about the specificities of their way of life. It is likely that it is precisely the fact of being a foreigner that makes this incursion possible, insofar as the tourist, unlike many locals, is not emotionally affected by neighborhood violence and, apart from the brief moment when they voluntarily venture there, does not share fears associated with the space and built up throughout their lives by a set of fictions, rumors and information disseminated on a local scale. Nevertheless, the tourist valorization of the favela is not without local impacts, and we can consider that the favela tours have, in some respects, given the impulse for local forms of valorization and new practices in the favelas, as evidenced by the frequentation by Cariocas and Brazilian tourists of the Bar da Laje at the top of Vidigal. However, it is the tourist who precedes the neighbor here, and we can consider the taste for marginal urban spaces as a Western privilege, slowly diffused on a global scale.

Indeed, a reading of the ethnographic literature seems to pave the way for an approximation of practices observed in various places around the globe, particularly in India (Meschkank 2011) and South Africa (Rolfes et al. 2009), and which some researchers have even grouped together under one category, slum tourism (Frenzel et al. 2012). In order to analyze the nature of tourism motivations for visiting favelas, it seems necessary to focus on developing a more general valorization of the urban margin, which would take into account more cross-disciplinary elements and less related to the purely site-specific contexts visited. In this regard, it is perhaps interesting to note the similarities between the different places where tourists visit the urban margins, in the main places considered by researchers as having relevant similarities (Rolfes 2010), such as Rio de Janeiro, Johannesburg or Cape Town and Bombay. The national contexts in which these different forms of tourism are embedded can be compared, as economists do when they speak of the BRICS, insofar as they are characterized by a recent phase of increasing neoliberalism and global power, even as they continue to display immense inequalities. These inequalities, often pointed out by the media when judging the economic aims of these nations, seem to impose themselves as part of the tourist landscape. The peri-urban area thus becomes a tourist site of interest – either in a more or less militant action, to discover the other side of the picture, the reverse side of the postcard, etc., which in any case is an integral part of the collective imagination, or that, one not excluding the other, these spaces have managed to appear as authentic places of the culture visited, in opposition to the modernization of the rest of the country. In this last perspective, the case study provided by Manfred Rolfes shows the proximity of the practices observed in Soweto and Rocinha, observing how the townships present themselves as the place of a rediscovered authentic Africanness (Rolfes, ibid.). In the context of townships, shanty towns or favelas, it is therefore an idea of authenticity that seems to emerge from the tourist imaginary and the promotion of practices.

6

From Exoticism to Authenticity

Although present nowadays in the relationship that the West maintains with otherness, the notion of authenticity is not a creation ex nihilo, but the result of a historical process. This history, that of the ways of thinking about the Other and their encounter in the West, is too long and too heterogeneous to be told. Indeed, the emergence of authenticity as a central value in the apprehension of otherness is caused not only by the development of tourism as a practice but also by the evolutions of the history of representations originally forged in the West. The most recent of these evolutions has led from exoticism, which dominated from the 17th century to 1900, to authenticity, the key value of the 20th and 21st centuries.

6.1. Exoticism

According to Salman Rushdie, "authenticity is the respectable child of old-fashioned exoticism" (Rushdie 1991, p. 67). This quote contains two essential elements. First, it assumes a connection between authenticity and exoticism, implying that the notion of authenticity, although new in appearance, is in reality the updating of an old notion, which thus hides the continuity of Western relations with otherness. I think, in fact, that it is only by tracing this connection between exoticism and authenticity that, recalling how it is neither objective nor completely relative, we can make these notions appear as they are, as the marks of a specific relation of the West to otherness. However, and this is the second element, it is necessary at all costs to guard against seeing in this connection only a continuity and not taking into account certain divisions. What Rushdie's idea also implies is the moral dimension of authenticity. The emergence of the concept of authenticity to conceal how exoticism responds to an ethical/moral challenge, to an imperative of legitimization that comes at a certain point in the history of relations between the West and the "rest" of the world. Part of the shifting from the regime of exoticism to that of authenticity can be assimilated to an enterprise of dissimulating the subjectivity contained in the exoticizing posture. Adorning itself with the bases of science, exoticism becomes authenticity and historically constructed relations are posed as objective. The rhetoric of authenticity stipulates that the conditions of difference concern something other than the point of view. It reifies otherness, positioning it as something scientifically definable, measurable and gradable. Authenticity leaves aside a set of otherness that is not different enough. In short, it is the euphemization, by science, of a unilateral relation in its attribution and in the very definition of culture.

6.1.1. *Exoticism, deictics and dialectics*

Exoticism, like authenticity, is not an objectifiable quality of objects, places or cultures. It is first of all a mental posture which, insofar as it directs the way Westerners act in the world, then gives rise to real practices, as well as to the production of new objects, places and images. A thing can only be said to be exotic in relation to another, and in the first place to something endotic. As Jean-François Staszak shows, exoticism as it developed from the 18th century onwards evokes the distant and the bizarre, adjectives "that seem to make sense in themselves, [but] do so only in relation to an implicit, relative to the speaker, his situation and his norms" (Staszak 2008). In this respect, as Staszak points out, all of the qualities covered by the field of exoticism correspond to what linguists call deictics: bizarre, distant, strange, in short exotic, are words that necessarily imply a position which, if not given explicitly in the discourse, contains the entire meaning necessary for understanding. Thus, exoticism can only be understood within a relation and on the condition that the transmitters of the discourse are designated. As it is constructed in the context of the Renaissance, and especially of colonization, exoticism is the fruit of the Western view of the world. Regarding deictics, the exoticism is thus also a dialectic that opposes the common to the bizarre, the near to the far and the normal to the extraordinary. For all of that, the terms of this relation are not interchangeable, and the oppositions tend to be reified in relation to the historical context. The golden age of exoticism coincides with that of the Western colonization and control of the world. If, from our point of view, exoticism is all relative, it is not perceived as such in a world where, as Staszak recalls, the cities of Paris and London became "absolute here" (ibid.). Exoticism clearly becomes the discourse of the European on the Other and, since it is part of a geography, of the temperate on the tropical.

This is an important feature, to which I will return, because authenticity pushes even further the principle that exoticism is not aware of its subjectivity and tends to establish a point of view into an objective reality. And it is thus obvious that to find the reasons for the development of exoticism in the modern era, it is necessary to look first to the West and not to the places to which it claims to turn: the fantasized tropics and the East. Two elements, concomitant, seem to be of equal importance here. The first is the feeling that the West acquires of its own power and the development of its ethnocentrism which, notably around the ideas of progress and evolution, will progressively establish its values as norms, its identity as humanity. The second is the development, in particular under the impulse of the romantic authors, of a critique of modernity and the rhetoric of nostalgia.

It should be recalled that exoticism is above all a literary theme which developed from a literature of travel, beginning properly only with the Renaissance (Moura 1998) and continuing with Romantics valuing the elsewhere. In the 19th century, it became, with Baudelaire in particular, the antidote to spleen (Mathè 1972). The veracity of what was described did not matter, and besides, the journeys were not necessarily real. On the contrary, some even put forward the idea that travel is precisely the cause of a destruction of otherness, and thus of exoticism. The idea is already found in Diderot's work (2006), as Anthony Pagden (1993) comments. For Jean-Marc Moura, exoticism is "a reverie which attaches itself to a distant space and is carried out in a text" (Moura 1992), and the literary context of its development corresponds to the emergence of a tendency, also literary, to criticize Western society that reached a climax in anti-modernism at the end of the 19th century. Moreover, for Staszak, if the devaluation of the Other correlates with the valorization of the self, then the reverse is probably true. But this devaluation of the self that the novelists operate is, of course, quite relative – rather lassitude than disgust (Staszak 2008) – as is the valorization of an Other that is above all dreamed of and fantasized. One of the natures of this fantasy is the vertiginous tropism of the Westerner who creates a confusion, that of time and space.

6.1.2. *Movement in space, travel in time*

As we have seen, the emergence of exoticism, which Moura describes as the step from the foreign to the strange, can be explained by a context in which European values are set up as standards. The 18th century thus corresponds to a first golden age of exoticism[1], which is reflected in all kinds of artistic productions. One work, *Bajazet* by Jean Racine, provides a good example of what exoticism did to theater. Certainly, the "turqueries" of Molière's *Bourgeois gentilhomme* had already amused the king's court, and there was an audience in France eager for exoticism. But *Bajazet* went further, breaking with the rules of classicism in two simultaneous ways: by setting the action in Turkey, a country too "modern" to be the setting for a tragedy and by adapting a story that Racine presented as recent and true. It was only in appearance that Racine broke with the rules inherited from antiquity for tragedy. And setting the action in Turkey was not (only) a means of pleasing the court and satisfying its penchant for exoticism, first satisfied in comedy. In fact, Racine

1 Although the word is attested only from 1866 (Mathè 1972).

explained the role played by distance in the preface to the work and showed that this alone justified his "audaciousness" to stage a story from his century, even though classical tragedy requires the ability to look at the characters on stage with a different eye. He wrote in 1676: "The distance of the countries repairs in a way the too great proximity of the times" (Racine 2017, p. 564). The time was not yet to represent everyday life, and the familiar had no place in the theater. And this is how exoticism seems to appear: as a renewal of the framework, which consists of a displacement of the temporal distance (which made antiquity the time of tragedy) towards a spatial distance, from which orientalism was born.

This is not an insignificant fact, and it refers to what Jean-Loup Amselle said about exoticism when he wrote that it operates a "dehistoricization" of the Other (Amselle 2010). The idea that the Other, the exotic Other, sharing the same time is denied because the distance that constitutes the relationship must always be conserved. The idea that Racine exposes, little consciously no doubt, in the preface of the only one of his plays that dares to stage his century, contains a confusion between temporality and spatiality that remains constitutive of the relations maintained by the West with the otherness. Evolutionism in anthropology would push this confusion to its paroxysm and will soon consider that a temporal distance necessarily corresponds to the spatial distance with the exotic Other. At the forefront of these theories and comments that flow in, we, of course, find the myth of the "good savage", dear to the philosophers of the Enlightenment. Although the scope of the philosophers, more explanatory, moved away in theory from the poetic aspirations of the artists, the myth of the good savage shared with the exoticism a certain number of capital traits, among which enthroned ethnocentrism. Ethnocentrism is a denial of the idea that so-called exotic people are the result of historical processes and perceives them as fixed. These groups of people can only truly enter into history on the condition that they come into contact with the West. Without this contact, it is as if they remain in a suspended time. This ethnocentrism seemed, in the 17th and 18th centuries, completely unthought of (and there remain traces of it in our recent history), strong of the changes that the Western society recognizes, guided in the way of progress and propelled to the center of the world. In that, the belief that Racine expresses in a possibility of finding a temporal distance by a spatial distancing is not astonishing and testifies to the appearance, since the 17th century, of a nostalgia which expresses itself in exoticism. It should be remembered that this exoticism is as much linked to a feeling of power among Western society as to the emergence of a self-criticism which will sometimes overvalue the other societies, in a logic not less ethnocentric. Evolutionism, more distant and with a more scientific vocation, gradually made this nostalgia disappear. By passing from exoticism to authenticity, we will find a way to account for it.

For the moment, it should be specified that this merging of the temporal distance over spatial distance is a tenacious mental phenomenon, whether the goal is positive or negative. The positive version of this tendency, which consists of a momentary valorization of the Other, is what we call exoticism. It is obvious that a part of international tourism is still developing over those foundations and that, in the distant Other, we sometimes seek a certain idea of the Same. In any case, authenticity resembles this. Authenticity, like exoticism, is a relation in space which maintains a certain idea of time. Symptomatic of our never-satisfied desires to find in the Other both an unreduced difference and both the Same. A trend of this time of centralization and classification, of nostalgia and poetry.

6.1.3. *Exoticism of the end and the end of exoticism*

Exoticism, arising in the 17th century, has become a lasting theme in art and literature, especially in France. Around 1900, exoticism reached a new peak, especially in terms of its artistic manifestations. Numerous authors became exoticists and filled the imagination of readers with tales of the far-away. Skilled at rendering incongruity, they formed above all a new class of professionals to discuss the elsewhere. It was the time of Maurice Barrès, Paul Claudel, Saint-John Perse and, reigning among the writer-travelers, Pierre Loti. The fact that their works are somewhat obsolete today only links them more to their time and only accentuates the following statement: exoticism was at a turning point in 1900 and was about to change. Among the works of the great poet-travelers, Victor Segalen's work is the best evidence of this change. His *Essai sur l'exotisme*, which has remained in draft form, is unique in its aim: to show what exoticism is, and above all, to say the threat which hangs on it. The perception of that threat is the common attribute of most of the authors that I have gathered here under the denomination of professionals of the elsewhere. The central stake is then the conservation of the differences (intrinsic quality of the exoticism) in the decrease of the distances that the exoticism involves in part for its consumption. The shrinking of the world, in fact, accelerated at an unprecedented rate around the year 1900, was first marked by a revolution in transportation. As far as the distant elsewhere was concerned, it seems the question was less the railway than the development of steamboats. People were suddenly able to travel the world faster, with less risk and greater comfort. Moreover, comfort was standardized, and cosmopolitan European style could be found every day a little more wherever they traveled. In 1900, Europe had already finished dividing the world[2].

Two things deserve to be separated here. On the one hand, the denunciation of a manufactured exoticism, a counterfeit of the elsewhere. On the other hand, the feeling of a disappearance of difference, which only reinforces the feeling of

2 The Berlin Conference took place in 1885.

nostalgia that has been central to exoticism since its inception. These two criticisms foreshadow the emergence of authenticity.

The first criticism that points to the commercialization of exoticism and its counterfeit has origins before 1900. Thus, Alphonse Daudet's *Tartarin of Tarascon* is a scathing satire. The criticism of cheap exoticism already existed when, more vehemently, Victor Segalen exposed his disgust when faced with those that he named the "procurers of the Diverse" questioning Pierre Loti, whom he considered as being part of "impressionist tourists". Exoticism, which he called a "sensation of the Divers", had to according to Segalen be redefined in the face of its commercialization:

> Above all, clear the ground. Throw overboard all that is misused and rancid the term of exoticism. To strip it of all its trappings: the palm tree and the camel; colonial helmet; black skins and yellow sun; and at the same time to get rid of all those who used them with naive loquacity (Segalen 1978, p. 41).

Segalen is not alone in criticizing the commercialization of exoticism and in fearing the effects of emerging tourism. Because behind its commercialization, it is the arrival of new consumers of exoticism launched on the planet which appeared to be a threat. Perhaps the denunciation of what is lost in the reduction of distance should only be seen as a poetic transcription of what is in fact the result of a privilege. Still, the marketing of exoticism not in the form of books but in the form of travel (in the wake of Thomas Cook) pointed to the end of exoticism for them. In a few decades, Europe finished appropriating the planet, and exoticism lost its splendor. The term, little by little, fell into disuse. According to Anaïs Fléchet, World War I dealt the fatal blow:

> It seems that the positive valence of exoticism has been undermined by the new representations of the world resulting from the First World War. At the time of the finite world and the disappearance of white on the maps, the term becomes synonymous with artificiality; it is associated with the qualifiers: easy, deceptive or "cheap rubbish" (Fléchet 2008, p. 20).

Exoticism has gradually dissolved into its representations and differences have been reduced. As it is for blanks on the map, it is true that there are hardly any societies that are not, if not mapped, associated with an imaginary, represented and therefore miniaturized.

The feeling of the disappearance, not only of exoticism but also of cultural difference, is also present in the literature of the beginning of the century. A few years after Segalen, Paul Morand exclaimed with horror: "We are moving toward: around the world for eighty francs". He added that man will live then his most miserable hours: "All that one said of the misery of man will appear really only the day when this cost will be reached" (Morand 1928, pp. 14–15). Loti, already, had condemned diversity:

> There will come a time when the Earth will be very boring to live on, when it will have been made the same from one end to the other, and when one will no longer be able to even try to travel to amuse oneself a little (Loti 1899, p. 2).

Cultural standardization and the proximity of the confines of the world worried writers of exoticism. On the contrary, it is certain that Loti was mistaken in announcing the end of travel and of the search for exoticism through travel, which then was only beginning and continues to this day. Segalen had also made the conservation of exoticism, already undermined, the heroic goal of modern adventure. But, it is important to underline that all, in their denunciation, creating poetry, and that the disappearance of cherished diversity, far from threatening their poetry, found it and supported it. Indeed, the feeling of the disappearance, the announced and inexorable death, and the transience of the being of whole societies were the inexhaustible sources of their poetry. Morand knew it well; it was he who said that he found in Loti's texts "the taste of death" dear to young people. It was indeed the feeling of loss that founded exoticism at its literary peak and at the same time announced its disappearance: the end of exoticism was also an exoticism of the end. Here particularly, exoticism prefigured the authenticity that was going to be attached to demonstrate the survival of difference and to demarcate the true difference of its representation. So much so that, of this poetry of death which contributed to exoticism, there remained something in authenticity, and more generally in the understanding of Western otherness. This is what Barbara Kirshenblatt-Gimblett has shown in particular, by affirming that museums, tourism and heritage converge towards the representation of cultural traditions as lost (Kirshenblatt-Gimblett 1998). This is largely the nature of collective representations of cultural diversity in the West. Described as threatened or already lost, its romanticism stems in large part from a sense of its inaccessibility. When we celebrate difference, we enjoy the feeling of its fleetingness.

6.2. Authenticity, a scientific exoticism?

In this chapter, I have tried to specify the notion of authenticity, essential to the understanding of tourist representations, particularly in the case of favela tourism. Authenticity, indeed, is a concept placed at the center of the study of tourism by the majority of the illustrious researchers from this field. However, the notion of authenticity is found exclusively in tourism. It is found, on several occasions, throughout the history of anthropology. Moreover, we can consider the discipline as one of the major sources of the invention of the concept of authenticity and its influence on the relationship with otherness.

6.2.1. *Anthropology and authenticity*

Indeed, it seems that ethnologists have often contributed to tourist representations of otherness, when these are based on criteria of authenticity. Anthropology, as a discipline with a strong comparative dimension that has taken culture as its object, has sometimes taken on the task of evaluating the authenticity of a culture, that is, its conformity to a presumed tradition or, in other words, its degree of resistance to the threats posed by acculturation, diversity, modernization and globalization to the integrity of culture. In terms of culture, it seems obvious that authenticity is a quality (first in the general sense, then in the sense of value) that falls on the side of so-called traditional societies, those that are the classical objects of anthropology. Also, it appears that authenticity is, from the beginning, closely associated with otherness as it is defined by the social sciences. This association can be interpreted through the range of qualifications proposed for the foreign societies defined by ethnology which, according to Gérard Lenclud (1996), has often succumbed to the temptation to reify the categories of the "us" and the "other", the otherness then forming a unit opposed to the Western societies. This oppositional character of the denomination of the societies studied gives a lot of information on the ideas that anthropology has contributed to constructing: primitive societies/civilized societies; simple societies/complex societies; traditional societies/modern societies. Thus, the authenticity possesses from the outset a greater proximity with what is opposed to Western society, modernity is opposed to the tradition at the same time as the complex is opposed to the simple. Indeed, is there not an obvious semantic proximity between authenticity, simplicity and conformity to the tradition?

Moreover, authenticity seems to be a quality that is lost if one of these oppositions is crossed. Also, the loss of authenticity often results from a complexification process (let us understand it as a complexification of the object of study, of which diversity or globalization form salient examples) or from a change in tradition (acculturation). It is thus understood that authenticity, which maintains an underlying relationship with all forms of categorization of otherness, constitutes truly the criterion according to which a society, a culture, can or cannot be the object of ethnology. From quality in the first sense, authenticity thus becomes a quality in the sense of value as it serves to determine whether such or such a culture is capable of teaching us a difference, carrying information of an anthropological dimension (in the sense of the ethno/anthropo distinction formulated by Lévi-Strauss) or if, conversely, it is in a state of loss of its authenticity that places it in a position where it would then only constitute a weak version of our own culture or that of another. Thus, the foreword that Malinowski gave to the *Argonauts of the Western Pacific* in 1922 testifies:

> Ethnology is in the sadly ludicrous, not to say tragic, position, that at the very moment when it begins to put its workshop in order, to forge its proper tools, to start ready for work on its appointed task, the material of its study melts away with hopeless rapidity. Just now, when the methods and aims of scientific field ethnology have taken shape, when men fully trained for the work have begun to travel into savage countries and study their inhabitants – these die away under our very eyes (Malinowski, 2005, XI).

While the category of authenticity can already be interpreted through the Great Divide (Goody 1979) operated by anthropology (i.e. as a consequence of this divide, as a criterion allowing it to be maintained), it has sometimes been explicitly used by some of the most famous anthropologists. Also, Edward Sapir, student of Franz Boas and successor of a tradition of thought which defined culture as the object of study of anthropology, opposed genuine cultures to spurious cultures. On the contrary, it is obvious that Sapir's thinking is accompanied by a certain value judgment, even primitivism, when he affirms that genuine cultures, which are the so-called primitive cultures, are societies that are harmonious in all respects in comparison with spurious societies (industrial societies), which are those in which the individual is an "insignificant fragment of the social organism" (Sapir 1967, p. 89). We find this idea in the work of Claude Lévi-Strauss. The final chapter of *Structural Anthropology* gives a striking example, when Lévi-Strauss affirms about writing:

> [Writing] has taken away something essential from humanity. These society's [uncivilized and unwritten] are, to a far greater degree than the others, based on personal relationships on concrete relations between individuals [...]. Our relations with one another are now only occasionally and fragmentarily based upon global experience, the concrete "apprehension" of one person by another. They are largely the result of a process of indirect reconstruction, through written documents. We are no longer linked to our past by an oral tradition which implies direct contact with others (storytellers, priests, wise men or elders), but by books amassed in libraries [...]. And we communicate with the immense majority of our contemporaries by all kinds of intermediaries – written documents or administrative machinery – which undoubtedly vastly extend our contacts but at the same time make those contacts somewhat "unauthentic" [...]. In the

> future, it may be recognized that anthropology's most important contribution to social science is to have introduced, if unknowingly, this fundamental distinction between two types of social existence: a way of life recognized at the outset as traditional and archaic and characteristic of "authentic" societies and a more modern form of existence, from which the first-named type is not absent but where groups that are not completely, or are imperfectly, "authentic" are organized within a much larger and specifically "unauthentic" system (Lévi-Strauss 1963, 365–368).

In a logic of preserving their object of study in the face of what could undoubtedly be perceived as a source of pollution and included in the processes of change, miscegenation, modernization, acculturation, etc., some great anthropologists have thus made authenticity an essential criterion, and one that serves in a way to fix the minimum degree of otherness necessary for the application of the ethnological and ethnographical method. As Handler and Linnekin (1984) have shown – in particular concerning the opposition made by Hobsbawm (Hobsbawm and Ranger 1983), recalling the one posed by Sapir between spurious traditions and genuine traditions – the perceptions of culture and tradition that are expressed through this importance given to authenticity are based on a naturalistic conception that needs to be overcome. Indeed, the criterion of authenticity only makes sense insofar as we conceive the possibility of finding an original and fixed model of the thing to which we apply it. These authors thus contribute to the diffusion of a somewhat essentialist definition of culture and tradition, and above all participate in the creation of a conceptual framework in which it becomes possible to graduate otherness and to evaluate, behind the criterion of authenticity, its value. It is very likely that these academic uses of the criterion of authenticity in the study of otherness have been transmitted, more or less directly, outside the university world. Beyond Lévi-Strauss's prediction, I try to develop here the idea that, if we can discuss the importance of this contribution of ethnology to the social sciences, this "capital distinction between two modalities of social existence" constitutes without a doubt its most striking contribution to common thought. By tourism, it is expressed indeed the reception of this methodical exoticism and its use of the authenticity as measurement.

6.2.2. *Uniformity of the world and authenticity*

The feeling of authenticity projected by tourists and staged by guides in the favelas is closely linked to the construction of a marked otherness and conditioned by the idea of a separation between the world of the tourist and that of the favela. The aim here is to clarify what, in the tourists' interest in the favelas, is not only the creation of an otherness (without which certain corners of the city, never visited by tourists, could arouse the same passions), but is akin to a valorization of it established in a moral perspective of self-criticism and exoticization.

Many authors have highlighted the existence of a process of moralization in contemporary tourism[3]. Among them, Jim Butcher (2003), an anthropologist of tourism specializing in development issues, has described contemporary tourism as a phenomenon characterized by the propagation within it of a morality evolving according to a double dynamic. On the one hand, this morality is based on the elaboration of a critique of the so-called "mass" tourism, which is pointed out as a danger to both cultures (which it threatens with inauthenticity) and to the environment. On the other hand, this contemporary moralization of tourism appears to be the driving force behind the advent of new forms of tourism, which are available under multiple forms and terms (solidarity, ethical, moral, cultural, etc.). In many respects, the moralization of tourism, which condemns certain practices while constructing new ones, bears witness to a reversal of the representations of Western modernity.

While, as Jean-Didier Urbain (1991) has shown, the pejoration of the tourist is an old phenomenon which, as the example of Flaubert for whom the term is already negative, goes back almost to the origins of contemporary tourism (i.e. to the Grand Tour), the diffusion of a negative perception of mass tourism is not entirely contained in it. In fact, it would be tempting to summarize any criticism of tourism as a matter of distinction. But, while they are indeed important driving forces of the diversification of tourism practices, they do not explain the entire evolution of our relationship with tourism. It would, in fact, be unfair to never consider as sincere the concerns, sometimes expressed in hatred, that have developed with regard to the expansion of tourism. A large part of the evolution of our theoretical relationship with tourism is linked to a certain inversion of our relationship with modernity as a project. If we look, for example, at the nature of Thomas Cook's project, to whom we attribute some of the first initiatives to increase the practice of tourism on a mass scale, we see that the idea of progress plays a central role. Tourism was seen as a vector for the modernization of the world, perceived as a positive mission. Through the intervention of the Parisian elites of the Touring Club (Bertho-Lavenir 1999; Boyer 1999), it was also tourism that served as a lever for the modernization of roads and the standardization of structures for welcoming travelers. At this point in history, tourism is therefore at the forefront of modernization. It is not this association that has fundamentally changed. Indeed, tourism is still perceived as a vector of standardization and expansion of Western modernity. On the contrary, our relationship with it and the symbolic framework in which we interpret its evolution have undergone profound transformations. By extension of what some writers like

3 See, in particular, the article by Nadège Chabloz on the creation of charters and codes of ethical conduct for tourism (Chabloz 2006).

Butcher have presented, we can put forward the hypothesis of an inversion of our relationship with modernity (and therefore of a certain growing dislike) as being at the origin of our critical relationship with tourism.

In this phenomenon, it is certain that the recognition of global warming in general, and the harmful ecological consequences of tourism in particular, has played an important role. It is likely, in fact, that the consideration of tourism as a danger to cultural diversity has expanded from the ecological to the cultural realm. In his *A Sand County Almanac*, Aldo Leopold, one of the fathers of American environmental ethics, writes:

> But all conservation of wildness is self-defeating, for to cherish we must see and fondle, and when enough have seen and fondled, there is no wilderness to left to cherish (Leopold 1949, p. 101).

This perception of irremediable pollution by man, which is paradoxical because it is through "cherishing" that he indulges in it, has the appearance of a prophecy which we can see is now often applied to tourism, and therefore to the field of culture. While tourism has real cultural consequences, if only because of the fundamental inequality that opposes the leisure mobility of the traveler to the constrained immobility of the receiver, it is their interpretation that is at issue here. Indeed, tourism is often perceived as a vector of standardization of the world. Using ecological lexicon, it is often described as a danger to diversity, not only environmental, but also cultural. Of course, a rather fixed perception of culture is at work here. In any case, it seems relevant to link the development of a critique of tourism to a progressive disassociation with the project of modernity and the disappearance, or loss of momentum, of the idea of progress. As I said with Butcher, this disruption of representations of the modernity/tourism pair must be seen as being at the origin, on the one hand, of the propagation of critiques of what is conceived as "mass tourism" and, on the other hand, of new practices of tourism. It is in this perspective that it seems possible to consider the development of tourism in the favelas and the construction of its authenticity as being largely based on its moral opposition to modernity and not only on the specification of its otherness.

7

The Favela in the Market of Otherness

The connection I wanted to trace between exoticism and authenticity supposes that, under different aspects, something common between these two forms of relationships with otherness has been preserved in the West. While I have sought to explain the reasons for the passing from one to the other, it is now necessary to make clear the continuity of the relations with the Other which, in the West, resists this change of paradigm. Responding to the requirements of different times and contexts, exoticism and authenticity appear then as two stages of a same construction process of otherness. They involve the staging "in" and "by" Western, modern and capitalist societies.

7.1. Authenticism and the crisis of otherness

Until now, this book has tried to show how favela tourism is achieved by questioning the roles of the rhetoric of authenticity. But, as I have indicated, authenticity is not an objective value, but a complex construction based on multiple oppositions. If it is not an objective value, authenticity may seem at times to be a value in itself, that is, as the object of the search undertaken by tourists through their visit. This impression may be perplexing, but it is based on the cross-cutting nature of authenticity, which is attached to all of the symbolic constructions observed, whether they are related to the reinforcement of the sensation of going to discover a real Brazil, a non-touristy, unknown world, or even poverty, authenticity being thought of as located in the interstices of modernity. In Chapter 6, I emphasized that authenticity, like exoticism in 1900, can only be fully understood in connection with the worrisome idea of a decrease in cultural diversity. While exoticism, discussed by Loti or later authors like Paul Morand, creates a poetic evocation of the disappearance of worlds and celebrates a form of otherness thought of as doomed to disappear, the regime of authenticity was about to impose itself. Under it, otherness must henceforth be regulated, proven and demarcated effectively from its manufactured avatars. The emergence of what I will now call "authenticism" is linked to this fear of the standardization of the world, which threatens otherness insofar as it is and remains a reason to travel. In this, authenticism and the final hours of the literary exoticism have in common the perception of a rarity and a crisis of otherness. For Marc Augé:

> The crisis of modernity, where some see a crisis of identity, could be rather attributed to the fact that one of the two languages (that of identity) prevails today over the other (that of otherness). It would thus be better described as a crisis of otherness (Augé 1994, p. 87).

The same is true of Marc Guillaume, who, in his book of interviews with Jean Baudrillard, writes:

> Western societies have, on the contrary, reduced the reality of the other through colonization or cultural assimilation. They have thus reduced what was radically heterogeneous, what was completely incommensurable in the other. So, in a world where a relative material abundance has arisen, we can say that the real scarcity is otherness. Perhaps then, the only way to fight against this scarcity is to invent a fiction of the other (Baudrillard and Guillaume 1992, p. 48).

The crisis, or the scarcity, of otherness could well be one of the foundations of the tourism practice in its cultural form, when it involves meeting an Other celebrated in its difference. In this, tourism is part of an economy of otherness. Saskia Cousin appoints the "economy of otherness" as the "staging and marketing of symbolic goods of objects, services, knowledge, and images, whose value depends on their character of otherness – that is, they are produced by others and/or for others" (Cousin 2018). Authenticity is a value intimately linked to this economy of otherness. Warnier, for example, has shown that modernity is based on a rhetoric that locates what is true in the elsewhere (Warnier 1994). The elsewhere and the Other are thus the holders of authenticity: we can draw from them goods of otherness, convertible into symbolic goods.

I have described here an essential aspect of the place taken by the favela in a market of symbolic goods. This aspect, the "rhetoric of authenticity", thus dominates the symbolic economic conversion of the favela. Authenticity is an essential value for symbolic economy. But, as we have explained in previous chapters, it is not a value in itself (although it tends to be established as such), but a quality of the object that is itself the symbolic good. The favela enters a symbolic market as a good of otherness. Authenticity acts as a support for otherness, which allows it to survive. The tourism that we observe in the favelas, in that, is not so much a quest for authenticity as a quest for otherness. The value of otherness, moreover, is common to exoticism and authenticity. Under the regime of exoticism, the value of goods depends on their otherness. This is also the case under the regime of authenticity, since it is first of all an opposition between the Same and the different, which places the authentic character on the side of the latter. The character of otherness, untainted by any pollution of the Same, fixes the value of tourist objects placed in a symbolic market. It is by consuming this otherness that travelers succeed in constituting their experience as a symbolic good that can reappear, back home, long after the visit. The three main axes on which the discourse of the guides and the post-visit narratives are developed are so many marks of authenticity of the favela's otherness. In the tourist framework, the favela is both a true Brazil, an otherness preserved by tourism, and a place of life for poor inhabitants, an otherness of urban interstices. Visiting them is both to experience for ourselves a rediscovered cultural diversity and to collect these symbols to enhance the positive symbolic charge of our mobility. Otherness is therefore relative. Ultimately, it exists only insofar as it can be incorporated into narratives of the world within which we can judge its authenticity. Here again, the discourse of otherness gives way to the discourse of identity. Since authenticism dominates, and because at its core it refuses the character of otherness to what comprises the Same, it is indeed from this latter that the definition of what Other really is, begins.

Above all, all of these experiences of otherness, skillfully constructed and reinforced in their tourist staging, are likely to be converted into symbolic capital. The scarcity of otherness, exacerbated by a discourse on the uniformity of the world, transforms the object of tourism into a symbolic good. I have already commented on the symbolic value of experiences in Chapter 5, where we discuss the strategies of distinction employed by tourists to valorize their experience in the favelas. The notions of a "market for symbolic goods" and an "economy of otherness" help to further explain why such strategies are particularly important in the context of favela tourism, as spaces with locations in the interstices would have allowed for the preservation of an otherness. It is because favelas are destined to take on a tourist value that stems from their otherness that a distinctive struggle emerges over which practices are deemed suitable or not for experiencing and then being able to narrate them. The rhetoric of exploration, of crossing borders and of adventure in the non-touristy areas, which characterizes the tourism of favelas, also explains the frequent recourse to an authenticist logic.

7.2. Miniature worlds

Tourism, when it has otherness as its object, often looks for spaces preserved from modernity and its standardizing effect. Even more than in the time of exoticism, the regime of authenticity, at work in the tourist valorization of the favelas, is built over the theme of disappearance. This time, disappearance is thought of in a universalist framework, which links the entire human species. Otherness is no longer an amusing curiosity, but the sign of the survival of other possible ways of life away from modernity. The fact that places and others come to embody them is important for them to be able to think about themselves, dream about themselves and experience themselves through the journey of other possible worlds. During the voyage, the Other becomes the model on which we think about our own identities and our "disalienation". The myth of the authentic Other joins that of the authentic Self (Selwyn 1996). Like exoticism, the regime of authenticity organizes a revision of time. If exoticism does this in a confused way by mistaking spatial distance for temporal distance, authenticity does so it in a stricter way. It is no longer a question, today, of going in search of the past in distant lands for which modernity will be the future for sure. Rather, authenticity delimits resistances to modern times and celebrates them for what is immemorial.

The object of tourism becomes a projection of what is considered to have disappeared in modern society and to be missing in the constitution of the Self. That amounts, as Pierre Nora said, to work "to the deciphering of what we are in the light of what we are not anymore" (Nora 1984, p. XXXIII). There, through what is different and not truly history, it is also a question of rediscovering the self, buried and polluted by modernity. In this perspective, it is something other than a Them and Us that seems to be fundamentally opposed in the practice of tourism. Indeed, we should not underestimate the importance of this relation to otherness in the construction of the self. Let us recall here the assertion of Segalen, who exclaims: "As always, we made a journey afar from what was only a journey in the depths of ourselves" (Segalen 1976).

David Picard (2011) proposed speaking of "human gardens". These consist of thinking of tourism as the action of cultivating spaces with an allegorical dimension. It is an action, undertaken by groups and individuals whose powers and interests diverge, which therefore leads to a negotiated situation, whose aim is, in the manner of the royal gardens of the *Ancien Régime*[1], to symbolize ideas, feelings or paradigms that are either by nature immaterial or are perceived as disappeared or distant. The image of the garden is not exhausted with the royal gardens, whose allegorical dimensions and "miniature world" aspect are very poignant. The garden, in fact, is also the space that we build between nature and culture, between the domestic and the unknown. Cultivating such places in this way will therefore allow us, by visiting them, to experience not really a place, but ideas, ideals and feelings. In the case discussed here, the favela is thus cultivated as the representation of spontaneity, of counter-modernity, the preserved authenticity of culture and social cohesion. The favela thus comes to materialize a set of ideas which symbolize a

1 The political and social system of the Kingdom of France from the Late Middle Ages until 1789.

possible alternative and come to counterbalance the painful sensation of the standardization of the world and of the continuous destruction of the worlds. The touristic favela has something of this, of the miniature world built to symbolize concepts that are too abstract or ideas that are too big: poverty and its culture, the real Brazil, the maintained otherness. This is how, perhaps, we can make tourism appear as a taming of the world, referring also to Bachelard who, echoing Schopenhauer writing that: "The world is my imagination" (Schopenhauer 2004), specifies: "I possess the world all the better as I am more skilful at miniaturizing it" (Bachelard 1957, p. 142).

In this perspective, tourist practices are thus approached as an act by which the individual traveler constructs and models his relationship with the world. Travel, in this sense, has something of the order of a ritual, an idea developed in particular by Nelson Graburn (1977), for whom tourism corresponds to other ritual institutions through which human beings work to embellish their lives and add meaning to them. By granting the act of travel, such a dimension thus makes it possible to go beyond the idea of tourism as simply being an escape from the everyday, thinking of it more as an intellectual project by which the setting in abeyance of this everyday constitutes a liminal moment through which it is not forgotten, but in a way set aside. In the same way, the stake of exoticism, or of authenticity, no longer appears only as that of a consumption and a saving entertainment, but rather as that of a marker of an otherness strong enough for this ritual of creating distance to be accomplished. In other words, the figure of the Other appears to be essential in the way of thinking or rethinking our own identities. Tourism, in this respect, presents itself as a particularly effective activity in this exercise of liminarity. By only taking into account the idea of confrontation with otherness as a project of construction of our own identities, tourism thus appears to be in continuity with principles that predate industrial modernization, and which were already expressed within travel. Thus, as Francis Affergan notes concerning André Thevet's travels in Brazil in the 16th century:

> Leaving, traveling, and discovering induce a pragmatic reversal of perspectives: one expects from an unknown, but often imaginary, otherness a contribution to the foundations of one's own origins (Affergan 1987, p. 105).

So, through the Other arises the Same. And if the Other is the medium of the journey, it is very often the Same that is finally sought, either because the quest is basically a quest for ourselves, for our own authenticity or identity, or because, all things considered, it is the Same in the Other that constitutes the unique horizon of intelligibility. Todorov (1989) has established a typology of travelers, dividing them into 10 types. Any typology of tourism and tourists, and there are many[2], should be taken with caution, as the different profiles of travelers are often mixed into one and the same person during the same trip. They are also often subject to extrapolations that are not unrelated to the desire to valorize one practice and/or devalue others. As a support for a theoretical exercise, and therefore accepting from the outset that it is a question of archetypes, they do have a certain interest. Todorov's interest is composed of ten archetypes: the assimilator (who travels to convert, in the image of the missionary), the profiteer (the merchant, for example, for whom travel is only a means), the tourist, the impressionist traveler (that of the traveler in the 19th century, seeking to enjoy difference), the exote (Segalen's perfect traveler, who saves distance and rejoices in it), the exile (a borderline case of the traveler), the allegorist (who sees the foreign tourist as a critical metaphor), the disillusioned (a variation of the allegorist, but who travels only to find that one is better off at home) and the philosophical traveler (like Montaigne). Starting from the object, the destination, it is impossible to say anything about the archetypal forms potentially attracted by it. Indeed, this typology does not say what the object is, but how it is consumed, its use. And in fact, it is quite possible that, in a case like tourism in the favelas, all forms of travelers potentially exist. Even a long fieldwork among tourists, such as the one that provides material for this book, does not allow us to conclude whether there is a particular trend. It is interesting to see the case of the tourist in this typology. After all, it would probably be enough to remove the *tourist* type from the typology to get the impression that it is precisely the tourist who is being used to construct the various archetypes. Only the category of the exote finally seems to be dubious and seems hard to be found in reality, beyond a figure developed by Segalen in an ideal way. The archetype of the profiteer is also of little relevance to our reflection. All of the others have in common that they approach the Other with the aim of a return to the Self; they are all allegorical in some way.

In this, favela tourism also appears as the result of a mental, collective and unconscious process, which involves making this space, itself fantasized in its homogeneity, come to embody values considered as disappeared or absent from Western societies. It would be superimposed on a real place, an imagined place, which could be qualified as heterotopia, as proposed Saskia Cousin (2020) or Alix Boirot (2020) from the definition that Michel Foucault gives, qualifying these spaces of "kinds of utopias effectively realized" (Foucault 2001, p. 1574). By the tourist heterotopia, a "space of illusion which denounces as more illusory still all the real space" (ibid., p. 1580), it is Western society that interprets itself mirrored. But, although the heterotopia is a projection of the imagination, it is the allegorical process that conditions the setting in the tourism of places. It is not less a mechanism of reduction of the otherness, which reifies it and structures it according to subjective principles.

---

2 See, in particular, Cohen (1972), Sharpley (1994) and Mehmetoglu (2004).

7.3. Otherness and tourism, between celebration and domestication

In *The Tourist*, a classic of the anthropology of tourism, Dean MacCannell gives a fundamental place to authenticity, which he summarizes as follows:

> Modernity first appears to everyone as it did to Lévi-Strauss[3], as disorganized fragments, alienating, wasteful, violent, superficial, unplanned, unstable, and inauthentic. On second examination, however, this appearance seems almost a mask, for beneath the disorderly exterior, modern society hides a firm resolve to establish itself on a worldwide base. Modern values are transcending the old divisions between the Communist East and the Capitalist West and between the "developed" and "third" worlds. The progress of modernity ("modernization") depends on its very sense of instability and inauthenticity. For moderns, reality and authenticity are thought to be elsewhere: in other historical periods and other cultures, in purer, simpler lifestyles. In other words, the concern of moderns for "naturalness", their nostalgia and their search for authenticity are not merely casual and somewhat decadent, though harmless, attachments to the souvenir of destroyed cultures and dead epochs. They are also components of the conquering spirit of modernity, the grounds of its unifying consciousness (MacCannell 1976, pp. 2–3).

Often summarized as the idea that tourists would always be attracted by authenticity, the theory that MacCannell exposes in *The Tourist* links tourism and modernity. For him, tourism and the representations of otherness on which it is based are the ideological expansion of modernity. To elaborate his theory, MacCannell relies on a structuralist perspective in order to describe the progression of modern society around two opposing but complementary processes, namely, differentiation and assimilation. Tourism is a complex game (a ritual for MacCannell) that constantly oscillates between these two processes. On the one hand, tourism is necessarily an enterprise of incorporating fragments into a unified experience. On the other hand, this enterprise is necessarily doomed for failure (and thus doomed to be repeated), since tourism celebrates difference. Georg Simmel has already defined this dual mechanism in a text entitled "Bridge and Door", in which he defines the movement by which society is constituted. The door symbolizes the aspect of social life that involves separating and generating differentiation: by creating doors that he can open and close, man produces division. The bridge, on the contrary, produces connection and cohesion. For Simmel, society is produced by

3 Dean MacCannell refers here to Claude Lévi-Strauss, who had presented, at the time of a speech MacCannell had attended in Paris, the idea that an ethnography of modernity was impossible because of its complexity and its fragmentation.

these two extremes of social interaction, and it is interesting to note that although the processes are apparently consubstantial, Simmel nevertheless asserts a definite order. Indeed, the door always precedes the bridge, which Simmel logically proclaims: "Things must first be separated from one another in order to be together" (Semmel 1994, p. 5). Of course, Simmel does not have in mind, unlike MacCannell, that such a functioning could be applied on a global scale. It is precisely here that it is interesting to take up what I said earlier about exoticism, associating its birth with that of a sense acquired by the West of being at the center of the world.

Exoticism and authenticity are similar in terms of processes of assimilation of diversity, which at the same time, celebrate difference and consume it, order it, classify it and also limit it. Exoticism in the literature, and in art more generally, has contributed to this vast operation of nomination, of the conquest of the world by the imaginary. Here, of course, this was not necessarily the intention of these artists, but their status had already transformed the elsewhere into a symbolic good. It is, as I have already mentioned, striking to observe the differences which separate exoticism and authenticity as for the mode of their evocation. Contrary to authenticity, which belongs discursively to the regime of evidence, exoticism is best evoked by a poetry, by the feeling and the impression. Faced with what he calls the "decline of the Diverse" and the acceleration of the process of modernizing the world around 1900, Victor Segalen advocates a vision of the difference which underlines the irreducibility of otherness. For him, the pleasure to enter in contact with the Diverse comes from an "acute and immediate perception of an eternal incomprehensibility" (ibid., p. 44). Otherness, to arouse exoticism, must thus be impossible to assimilate. It is indeed against any assimilation that he writes: "Let us not flatter ourselves to assimilate the manners, races, nations, others; but let us rejoice to never be able to do it" (Segalen 1978, p. 44). The exotic ideal, for Segalen, is an assumed subjectivity that remains an impression, a taste, a feeling. But Segalen is a late "exote", especially worried about the disappearance of the differences, on which he develops a poetry marked by nostalgia. Nowhere else is the deconstruction of the exotic feeling and the affirmation of its irreducible subjectivity more evident than in the *Essai sur l'exotisme*. Most of the other exotic authors, since the 18th century, have kept this relation unconscious and make of exoticism a value in itself, denying the evidence of the relational character, and thus all relative of the feeling. Also, for Jean-François Staszak, "the anonymous discourse that one does not situate carries the mark or rather the mask of the neutrality, which turns out in fact to be that of the dominant discourse, that of the white man" (Staszak 2008). And, above all, what hides behind the apparent objectivity of exoticism is a selection and division between good and bad difference. Exoticism is first and foremost concerned with what allows for a poetic or trivial evocation of what is different, bizarre and capable of evoking nostalgia and the beauty of distant lands. It effectively excludes the changes seen as too radical: "Is exotic only a measured strangeness, acceptable, apprehensible. Domesticable and domesticated. The exoticism is pleasant, it should not frighten or question" (ibid.). It is there, in this selectivity, that the sign of an enterprise of assimilating otherness is seen.

Peter Mason distinguishes two stages in the exoticization process that show how exoticism is an assimilation of otherness. First, "the exotic is produced by a process of decontextualization" (Mason 1998, p. 3). This process of decontextualization is followed by a recontextualization of the element, which takes on new meaning in the society in which it is newly assimilated and where it has become exotic. This is notably the case for objects that range from ritual use to ornamental status, or for goods, foodstuffs, for example, that find new practices. The place of objects in exoticism is crucial. Moreover, Anaïs Fléchet (2008) reminds us that the word "exotic" appears for the first time in Rabelais' writings to describe goods.

Exoticism thus evolves in and through the negation of a unilateral relation that presides over its application to different kinds of objects. Few authors, like Segalen, make subjectivity a conscious principle. On this point, it is obvious that the idea of authenticity goes further. The otherness described and assimilated (notably by several decades of ethnology) is quantifiable, more or less different or more or less polluted by modernity. Authenticity is perceived as objective; it does not depend on the point of view. Moreover, exoticism has a particularity: the term can apply to the exotic object as well as to the feeling of its encounter or its evocation. Nowhere do we speak of authenticism designating the tendency to value and to seek an intact otherness. By the very fact of its episteme, which it borrows from science, authenticity conceals its relational and subjective character. By showing itself as a potentially objective value, authenticity operates, much more effectively than exoticism, the euphemization of the unilateral relation which presides over the definition and the classification of cultures and places. It conceals that it is built only on oppositions constructed mainly between images and experiences. Without doubt, in fact, the passage to the regime of authenticity has prolonged the assimilation of otherness in the modernity that had carried, for a time, the exoticism.

What is left for the other? Exoticism, as we have seen (with Mason, in particular), is an appropriation; the exotic is the result of a process that tears an object from its context and reincorporates it in a new one. Authenticity seems to have radical otherness as its horizon. We see this through the pronounced aversion of the seekers of authenticity for all forms of pollution, for all that could have derived is a culture of a supposed state of its own and individualized cultural identity towards a state where it would have undergone the influence of other cultural forms, in particular those of the Same. But, besides the fact that the idea of stable and distinct cultures is scientifically invalid, authenticity is itself a process that, while reifying (in theory), reduces and classifies difference. Radical, authentic otherness cannot be encountered anywhere, not because it does not exist or has not existed, but because the symbolic structures that command the apprehension of otherness in the

West always push to reduce it, to order it, to include it in a system. This is what Baudrillard and Guillaume notably express:

> To put it simply, in every other there is the other – that which is not me, that which is different from me, but which I can understand, even assimilate – and there is also a radical, unassimilable, incomprehensible and even unthinkable otherness. And Western thought never ceases to take the other for the other, to *reduce* someone else to the other (Baudrillard and Guillaume 1992, p. 10).

This transformation of someone else into the others echoes, of course, the shift from the foreign to the strange, which Mora sees as the sign of the birth of exoticism. Authenticity plays a role in allegory as it does in impressionism, as it does in almost all forms of travelers in Todorov's typology. Does exotism? Indeed, except that authenticity is a viewpoint that claims a radical otherness that we cannot potentially judge. But since it seems that remaining in this state of intelligibility is impossible, it is also a quality attributed according to the coherence of the Other to an image to which he is supposed to correspond according to the (quite Western) values that we expect him to embody, or that he already embodies if the process of exoticization has passed. In this, we can see the connection that exists between the two, with exoticization, far from having disappeared, being an earlier process than authenticity.

His interest in the other, due to the fact that he is exotic, supposes that it is because we find in him a domesticated, selected and pleasant otherness. To be interested in the other by seeking his authenticity is to attest to his difference. Here, the process does not claim its goal explicitly. Obviously, I have largely referred to exoticism and authenticism as processes of assimilation that consist at least of including the Other in a space of intelligibility, so that it makes sense for ourselves, that is, the Western self. Exoticism and authenticism both have processes entirely linked to the incorporation of the distant in the near, of someone else in others, of the stranger in the strange, of the Other in the Same. There remains, however, less of a higher stratum of motivations to these processes which is not elucidated and of which the assimilation is the effect, not objectively the goal. In any case, a goal closer to the aspirations of those who engage in these discoveries of the Other should be sought, since assimilation cannot be considered a conscious objective, especially since from the moment one enters the regime of authenticity, it is perceived as antinomic.

After having seen how the favela is formalized in an economy of symbolic goods, it remains to make a return to the field, in order to observe how practices develop and how the meeting between practices and their object is constructed under the regime of authentic otherness. This return to the field will highlight certain logics of avoidance practiced between the inhabitants and the tourists, which show that several points of view are confronted which do not assimilate. The favelas that tourists see are perceived in a regime of otherness and in an economy of symbolic goods that does not include everyone as actors. The favela inhabitants continue to experience the favela according to a regime of identity that corresponds to its own symbolic stakes. This identity (like otherness) stays as it always was: a moving, evolving identity, which is fixed by places and undone by others, without ever being fixed. At times, tourism practices clash these modes of perception and appropriation of what constitutes the favela. They clash for the most part without violence, so much so that they cross each other, touch each other without ever penetrating each other. Only some experience the exchange at the price of symbolic violence, of light schizophrenia.

But authenticity is not entirely constructed upstream of the visits; it is not simply a mental construction exclusive to tourists, but is co-produced by the various actors at the time of the visit. Indeed, when tourism, as is the case with these practices, is centered on the recognition of authenticity, the different actors evolve in a certain tension. On the one hand, the tourists consider a large part of the authenticity of the place as being assured by the very fact of their absence and by the independence of the inhabitants with regard to their arrival. On the other hand, locals are well aware that the authenticity of tourists is different from the way they conceive their identity and leave the satisfaction of their curiosity to the guide. In some cases, the tourist situation is the result of an ongoing negotiation between tourists and the welcoming population. As a result, tourist representations tend to settle in a space between the identity prescribed by visitors and the identity claimed by receivers. But, as I will show, favela tours do not establish any such negotiation outside the control of the guide, and no real encounter takes place between tourists and inhabitants.

8

Gazes

I have presented the symbolic foundations of favela tourism, placed under the sign of authentic otherness. From now on, it is a question of analyzing how this symbolism takes shape during excursions, and thus questioning its structuring role in the contents and the course of the visits. For this, it is necessary to look beyond the discourses, which we have already seen, concerned the authenticist rhetoric of otherness, and focus on the way in which these discourses are articulated with the various elements that constitute the reality of the favelas. Finally, starting from the observation of an almost systematic absence of encounter between tourists and inhabitants, it is a question of evaluating the links between tourist symbolism and the interactional space in its observable way.

The lack of encounter is the first problematic element for an ethnography of tourist interaction in the favelas. In order to overcome this, it is necessary to describe a non-verbal type of interaction, best represented by the gazes that the different protagonists exchange or avoid during the visits. Several things can be said about these gazes. First, an order of interaction is detectable on the condition of observing how the gaze, often perceived as the free sense, is in fact conditioned by explicit or tacit norms and varies according to the situation. Moreover, it is a question of seeing in this interactional order, about which I will be led to speak about a normed space of the gaze, how it itself partly conditions the construction of authenticity at the heart of the tourists' experience.

Dealing with the gaze in anthropology requires a specific ethnography. It is a question of observing and noting not only what individuals do and say but also what they look at, and when and how they look. In the case of the fieldwork, which serves as the starting point for the reflection in this book, the gaze has imposed itself spontaneously as a central point of the observations, because it makes it possible to fill in two absences that are damaging for the understanding of tourist practices. The first is linked to the profound nature of the tourist excursion, which in many respects is a primarily visual experience, even though the conditions of this visual experience are rarely the subject of direct work by ethnologists and sociologists. For these tourists who walk through the favela accompanied by a guide, the experience cannot be reduced to hearing a speech, nor to the few culinary or musical experiences that their hosts sometimes plan for them. If it is worth traveling to, it is for the opportunity to "see with your own eyes" this marginalized space, so long excluded from the world of tourists. Sight can thus be considered as the sense of predilection of the tourist experience. During the tours, sight fulfills at least two key functions: it transforms the experience in images on which a memory will be able to be constituted and it presents itself as the free sense, the one thanks to which we can develop a point of view independent of the discourse imposed by the guide. In other words, sight is conceived as the sense through which the visit is personalized, and as the means of exercising critical discernment (it is the idea that what is seen cannot lie), which makes it an essential weapon in tourist situations. When studying these situations, the second absence that an ethnography of the gazes comes to fill is that of the direct verbal interactions between the tourists and the inhabitants of the favela. Although announced by many tourist promoters, the encounter between inhabitants and travelers does not really take place, and the rare words that they can sometimes throw at each other percolate irremediably through the filter of the guide. The lack of direct interaction is primarily caused by linguistic incompatibility (bilinguals, whether tourists or locals, are more likely to find the means for direct interaction). But this is reinforced by the nature of the discourse, and more precisely of the general narrative that the guides construct in the favela. The authenticist logic that serves as a framework for the narrative limits the participation of inhabitants who are unsure of the role they should play in it.

8.1. Describing the gazes

Numerous difficulties, both methodological and conceptual, emerge when we aim to proceed with an anthropology of the gaze. These difficulties arise first of all in the field where the gazes are not always observable. Moreover, it is impossible to observe and then evoke a gaze alone, since there are always at least two gazes: that of the ethnographer and that of the observed. This irreducible impossibility of conceiving the gaze of others without using our own gaze has further consequences in the continuation of the scientific path, which involves describing in order to analyze. There, the gaze is not only a gaze that is looked at, but it is as if it is torn away from its existence, to the point that there is between a gaze and its evocation a difference even greater than between a landscape and its photograph. This is what Michel de Certeau expresses when he writes, "There is no possible nomination of the gaze: to name the gaze is already to narrativize it" (de Certeau 1985, p. 11). From then on, these reasons should be enough to relegate the gaze to social science oblivion – in fact, the gaze has remained a marginal object. A few sociologists have, however, sketched a sociology of the gaze in an approach comparable to the one defended in this chapter. In addition, an issue of the journal *Communications* (Haroche and Vigarello 2004) brings together a number of works along this line, whose conceptual history is presented by Claudine Haroche (Haroche 2004).

We can trace this history back to Marcel Mauss, who, in his famous *Techniques of the Body* (1973), dedicates some lines to the gaze and shows how it is the fruit of cultural learning. He discusses the differences in the use of the gaze: "Thus we should attribute different values to the act of staring fixedly: a symbol of politeness in the army, and of rudeness in everyday life" (Mauss 1973, p. 76). For his part, Georg Simmel (1999) comments on the exchange of gazes between anonymous urbanites and notes the importance of the gaze in the modern city; more generally, in the development of mass society. Norbert Elias, too, gives the gaze a role in the evolution of society. According to him, progress favors the gaze, a kind of passive relation of individuals in the public space, and moves away from all possibilities of physical contact:

> The pleasures of the eyes and ears are becoming ever more intense, richer, more subtle, and more widespread, than the pleasures of the limbs are ever more restricted by commands and prohibitions (Élias 1991).[1]

Mauss, Simmel and Elias come together first by the fact that they point out the social and cultural dimensions, and thus operate an essential movement which consists of extracting the exclusivity of the analysis of the gaze from the bosom of physiology or psychology. But their convergence goes further, insofar as all of them evoke the gaze in a similar framework, which is that of everyday life, and insist first of all on a possible social normalization of the gaze. At stake in their evocation of the gaze is indeed a social relationship that is engaged between several individuals, which refers to behaviors given and interpreted in the interaction. The gaze here is part of the communication system and constitutes an active principle of social interactions. In the same way, it is in this perspective that I place the reflections that follow, thus leading to consider the heritage of American sociology and the analysis of daily interactions, and thus clarifying a framework for the analysis of the gaze. Indeed, a link established from Mauss to Garfinkel allows us first to consider that the gaze is part of the domain of accountability. Harold Garfinkel coined the concept of accountability (Garfinkel 1967) to account for the ways in which individuals, during interactions, seek to make their actions understandable and legitimate in the eyes of others. In our perspective, the gaze can be first of all a means of showing ourselves to be conforming to the norms (intelligible and legitimate): for example, in a discussion, looking at the person who is speaking/to whom we are speaking can be a mark of our attention and respect, without which the attitude would be perceived

---

1 Author's translation.

as rude. Moreover, the gaze is itself subject to the requirement of accountability, and is not simply the mark of the intelligibility of our behavior. Thus, a gaze can be misplaced; it can be deployed in a way that is not adapted to the norms of the situation. The merit of the notion of accountability lies in this idea that through our attitudes subject to the gaze of others, we must also be accountable. Another example is found in the sociology of Erving Goffman, which also allows us to grasp the role of the gaze in interactions. In a social world perceived as a theater where individuals put on a show in order to "keep face", the gaze of others is a permanent threat, and the rule is to avert it, out of "civil inattention", when individuals find themselves in a disadvantageous position. Thus, gazes are themselves part of our normalized behavior and have an existence as phenomena of social interactions.

I thus take my approach to the gaze to complement a series of works which, from Mauss to Goffman, question the social and communicative modalities of the gaze. The place of the gaze, here, is thus different from that given by John Urry in *The Tourist Gaze* (1991), and which concerns a Foucaldian approach to the gaze. It is necessary to distinguish the approach defended here from a tendency to qualify by the term "gaze" a set of mental characteristics, whose framework goes far beyond that of their immanence within observable and describable social situations. Indeed, the analysis of the gaze has involved, in this movement, the examination of the relationships between symbolic structures and deployment of the gaze as main axis, while leaving to this deployment and to its description a marginal place. In doing so, the term "gaze" has sometimes become synonymous with what is supposed to underlie it. We speak of the gaze to evoke what happens behind the eyes, rarely to comment on what happens in front of them, that is, when gazes are exchanged in a given social situation. This perspective, from which this chapter intends to distance itself, is mainly due to an approach initiated by Michel Foucault.

In *Naissance de la clinique* (The Birth of the Clinic, originally published in French in 1963) Foucault develops the concept of the "clinical gaze", which characterizes the appearance of a medical gaze operating a detachment regarding bodies. The gaze in question here is a way of cutting up reality. It is this approach that John Urry took up in the field of tourism studies. Urry describes the construction of a tourist gaze on the world, which makes a selection, distinguishes elements to be seen and others to be concealed. Here, he points to the social construction of the gaze deployed through a filter of ideas, skills, desires and expectations. In summary, from all attitudes relative to class, age, gender, etc., Urry defines the gaze as opposing the vision, as it refers to "discursive determinations" and to what he names "scopic regimes". If I deviate, here, from this definition of the gaze, it is not, however, a question of calling into question his work, and two reservations must be formulated. First, my approach does not invalidate the Foucauldian approach, but is situated below or in parallel with it. The gaze has indeed a dimension that cannot reach its strict description, but it is indeed this one that fixes the ideal framework of this reflection. Second, his work evokes in places points that are relevant to my perspective. For example, Foucault, commenting on Bentham, evokes experienced dimensions of the gaze, induced by the panoptic system. To live in a panoptic system is to live with a gaze potentially permanently placed on ourselves. In this system, potentiality is equivalent to permanence (Foucault 1976), and I will evoke this idea in turn when, at the time of discussing what changes the fact of being situated under the potential gaze of others, I will evoke the Sartrean perception of the gaze.

The Foucauldian approach of the gaze is not sufficient if it does not describe the gazes how they are and how as they are deployed. According to it, the gaze appears as a container whose content concerns thought, representations, etc. The replacement of a reference to thought by the Foucauldian gaze is, in the strict framework fixed here, only a metonymic phenomenon. The Foucauldian gaze (later transformed to tourist gaze, male gaze, bourgeois gaze, etc.) designates a way of representing things, of constructing a perceptual framework that has effects on the presentation of reality, that makes a visual selection in reality, arranges it according to its perspective. In other words, this approach no longer affects the gaze in the act of looking, the verbal form is absent and the subject is no longer "the set of attitudes permitted or not, natural or not" (Mauss 1936). Our perspective focuses on the gaze as a structuring and structured element of social relations. It is interested in the gazes themselves and in their modalities of deployment in space, according to given social situations and described as they are directly experienced. In that, because it undertakes to describe the gazes by treating them first of all independently of the structures of representations which condition them. In each case commented on, it will be a question of real gazes that it will be a question of describing. This approach is therefore far from Urry's approach, inspired by Foucault, and closer to a sociology of interaction which is based on the description of the gaze.

8.2. The interactional norms of the gaze

In order to describe what conditions the type of interaction between tourists and inhabitants, and because verbal interaction is almost non-existent, it appears that a description of gazes is indispensable. Indeed, the description of the gazes gives first an idea of the rules which govern the deployment of gazes. If these are thought as free by nature, it appears, however, that there are constraints, standards and rules which condition them. The interest of the tourists, what they consider as the privileged object of their gaze during the visit to the favelas, is thus not the only element to take into account. The joint presence of the inhabitants and the tourists, whose position is understood here as an intrusion, dictates the limits of the gaze itself. As we shall see, this interactional limitation of the gaze is not without impact on the way in which tourists situate the authenticity of the object of their visit.

8.2.1. *Photographs and norms*

Before approaching gazes directly and specifically, a detour towards the practice of photography during the favela tours can allow a first entry into the world of visual norms. As a gaze materialized through an object (the camera or the smartphone), photography reveals the existence of norms that are specific to it but which, as I will show, are easily applicable to the gaze itself. In order to comment on photographic norms, it is worth keeping in mind that, when visiting favelas, ethical and security issues affect the practice.

Tourists are, in a way, subject to two obligations: that of being cautious in a context where they perceive themselves as potential prey, and that of being respectful of an ethical code that their presence transgresses at first glance, in that they are practitioners of international leisure mobility, in an underprivileged space traversed by immobile individuals on the scale of tourists. This situation, which makes tourists more cautious than elsewhere about their actions, is not without consequences for the way in which they concretely deploy their gaze. Security and ethical concerns also pose some challenges for the promoters, and it is their responsibility to assure tourists that they are welcomed and that their visit can be made freely, in complete safety, with the only condition that they be accompanied by a guide. Of course, there are no instructions about the gaze in the presentations of the different tour operators, who, on the other hand, make a point of reminding people that it is possible to take their cameras – this is not insignificant if we consider how much tourists are told not to walk around with them, wherever they are in Rio de Janeiro, including in the most touristy places. However, if the intrusion of tourists in the favelas can be synonymous of danger, if it can disturb on an ethical level, it is because it represents the intervention of an external gaze. Also, while the photographs are free in principle, prohibitions appear as soon as the tour of the favelas begins.

As soon as the guides are met, they give the tourists instructions about what they can and cannot photograph. The instructions given by Luis, a guide for the pioneering company Favela Tour, were quite significant. On the bus, before entering the favela, he took a moment each time to explain:

> You're welcome here, but there's one thing you have to know: don't take pictures of people without asking, don't take pictures of the inside of the houses, and don't take pictures of the traffickers.

Depending on the tour operator, the type of tour and the willingness of the guides to mention the issue of trafficking or, on the contrary, to avoid it altogether, the call for caution on photography is more or less centered on the relationship with the criminality of the place. For Obi, for example, who was born in the favela and opens his home up to the groups he takes to the favela, the presence of traffickers is the only reason mentioned as problematic for the tourists' photographs. He only said, "When I tell you to put the cameras away, you put them away", which he did several times throughout the tour. On the contrary, for Zezinho, who hates to talk about subjects related to violence, but who is always very worried about the perception that the inhabitants could have of his tour, the instruction is only ethical: "Don't take close-ups of people; imagine if they did it to you, it's very rude". In any case, the act of photographing is immediately posed as problematic and as involving a triple danger, which weighs at once on the one who photographs (and who is therefore likely to be called to account, sometimes suddenly), on the one photographed and on the guides as intermediaries responsible for the encounter.

The instructions concerning photography are generally respected. When the guide occasionally warns of a danger or a clear breach of ethics, their authority is obvious and tourists do put their cameras away, sometimes taking them off their necks and putting them into their backpacks. Conversely, the guide has the authority to point to an unmistakable object for photography; he points to it and tourists usually comply. However, a few incidents do occur. Only on rare occasions do the guides have to worry about one or two tourists who seem to have difficulty restraining their urge to photograph everything they see. It is especially when the group is walking through the alleys of the favela that it becomes necessary for the guides to check that a latecomer is not photographing the inside of a house. Sometimes, in fact, tourists hide from the guide's view to "steal" a photograph. In general, the local perception of the presence of tourists as a problem belongs mainly to the first-floor residents, who frequently find themselves face to face with tourists who may find it difficult to resist the temptation to take a look at their home or, worse, take a photograph. Yet, as we have seen, guides usually support the prohibition of photography of the interior of the houses. For Marcelo, guide and director of a large tour operator, this is a fundamental rule. During one of our tours, he had to reprimand a straggling visitor with a wandering eye:

> Most, 98 percent of the tourists, if you tell them, "no pictures", they say, "OK, got it". With the others, I don't always know what to do. Truly, if they don't understand that a tourist coming into your house, opening your door, taking a picture of you as someone typical of France, it is not on, then I don't know what to do. You don't go and take pictures of a guy like that, in his house and without asking! It's a question of respect. Today, the guy I talked to, I don't know if he was going to take a picture, but when we started to go up the stairs, I saw that he was not following. And in this little street, there was an open window, it's the house of a woman that we know well and who sometimes makes us waterproof ponchos. Sometimes she has her window open, her food is there. There, the guy stopped in front,

> looked at her, and looked at his camera. And immediately, I said, "no photos" and he rejoined the group (Marcelo, interview, July 2015).

Most incidents are avoided because it is a part of the guide's job to defuse such situations. The taking of pictures by tourists is the extreme case of this intrusion, but the gaze itself is also problematic, and the inhabitants often complain about it. This is reflected in this discussion with two women in their sixties, residents of Rocinha:

> Marta: Last time, there was one looking out my window [the other woman laughs as she shakes her head]. I said, "What do you want?" He looked at me and left [bursting out laughing in turn]. What is he looking at in my house? There's nothing to see!
>
> Luisa: [Laughs] Look, I have nothing against tourists, really. But for them to come into the house, then no. That's too much.
>
> Me: Don't you think it's also the guide's responsibility?
>
> Marta: Yes, possibly. But it's the same with the guides, I don't know them. They pass by, they say hello, they go away. But look, the street is for everyone. But the house is private (Conversation in the favela of Rocinha, June 2012).

Most of the time, a principle of avoidance prevails in these situations and no conflicts break out between tourists and locals, with the guide remaining the systematic intermediary. This is not the case with incidents involving drug traffickers, which represent all of the violent conflicts that have occurred during tourist excursions. In fact, the few incidents that have occurred in the 25-year history of favela tourism have been mostly related to photographs taken of drug traffickers. These incidents, often, are solved by deleting the photographs. These cases of transgression of the instructions are much rarer than in the case of the interiors of houses.

8.2.2. *From the photograph to the gaze*

In the examples I have cited, the gaze is not directly addressed as a potential problem, and the rules concern photography above all. However, and this is what I will try to show, it is certain that what applies to photography applies to the gaze, without this being stated by the guides and without this appearing to infringe a fundamental freedom. Between the prohibitions that concern photography and those that concern the gaze, the difference is thus one of degree and not of nature. It is obvious that the act of photographing is problematic, insofar as it creates an image whose diffusion escapes the object of the photograph (it is particularly true in the cases of confrontation with the drug traffickers, of which we easily understand the mistrust). However, we can look for other implications of photographs that are not relative to its purpose, but to the act of photographing itself. Now, in order to speak about these implications, it seems relevant to come back to what has been mentioned above, that is to consider the act of photographing and that of gazing through their similarities. Let us observe again, from this point of view, the instructions given by the guides. In reality, the photographic prohibitions also condition the possibilities of the gaze. What is forbidden to photograph is often something that is also delicate to gaze at, at least with intensity. In the case of the tourist who, because he was looking inside a house, was rebuked by Marcelo, who shouted at him, “No photos!”, the tourist immediately replied, “No, no, I didn’t take a photo”. However, there is no doubt that guilt, of intention of at least, is established in the situation. Here, then, it is indeed the gaze that is problematic, even though it is on the photo that the discourse is focused. In a way, the reference to transgressive photography, even though the transgression does not take place, occurs in a context where there has been a transgression of the gaze, something that is more difficult to reproach since the gaze can be presented as unintentional and, moreover, is perceived to be absolutely free. What photography has more than the gaze is to be more visible to others, to possess a materiality that makes the direction of the gaze undeniable.

The act of photographing is nevertheless a particular type of interaction with the world, which says something about the modalities of gazing in a given situation. Such a connection is already induced by Edward Bruner in his critical article on the film *Cannibal Tours* (Bruner 1989). Bruner poses several hypotheses, with the aim of establishing the functions of the act of photographing in a tourist situation. Of course, it is obvious that a significant number of the functions attached to photography reside in the materiality of its rendering, the printed photograph (or today’s digitized form, posted on social networks, etc.). Like the object brought back from a trip, the photograph consolidates the memory and materializes the moment when it was taken. Bruner notes in this regard that the photograph as well as the object, when shown upon return from a trip, do not really possess value in themselves, but serve more as material to support a narrative about the context of their realization/acquisition. But this is another function of photography, evoked by Bruner, on which some observations can be made. Indeed, Bruner’s article also defines the action of photographing as having an immediate value, which therefore requires the act to be detached from its productive purpose. He writes:

> The camera serves as a protection for the tourist/photographer; it puts them in a position of social isolation so they don’t have to interact with New Guineans face to face, eye to eye. They can hide behind the lens of their camera, which is a great tool for closet voyeurs because they can look, and even stare, without embarrassment (Bruner, ibid.).

The idea here is to consider that photographs are, consciously or not, taken for something other than to become objects. In other words, what is important in photography is not so much, or at least not exclusively, linked to its material destiny, but holds, perhaps above all in certain cases, the influence that the act of photographing has on our way of gazing *hic et nunc*. From this, several hypotheses on the functions of photography in a tourist situation can emerge. It is even quite simple to make a link between the situation described in *Cannibal Tours* and that of the favelas, insofar as both situations are likely to raise ethical-moral concerns. First of all, photography is an effective way to get rid of the impression of idleness, to escape contemplation. It is an act that gives the tourist the possibility of giving their gaze an active dimension, thus protecting themselves from accusations of contemplation (a protection that is particularly necessary in such situations), and perhaps simply preserving themselves from a certain boredom. When we photograph, we cease to be only a spectator to occupy an active position. Then, the camera enables a distance with the gaze of others: it puts the Other in the position as the object, while placing the camera in an intermediary position, and not its gaze and eyes. In this, the device allows its owner to extract themselves temporarily from the situation by making themselves unavailable for interaction, notably by concealing their emotions and abstaining from any reactions, or rather by limiting them to the movements of their device. Conversely, the act of photographing can make it possible to mark our attention and to conform, in the eyes of all, to the injunctions to look precisely at something on the instructions of the guide. In any case, taking a photograph is equivalent to doing something in a concrete situation, and it is an act whose implications are immediately interpreted. Nevertheless, this type of hypothesis is difficult to produce in a general tone and must above all invite us to focus on each field on the present and concrete dimensions of the act of photographing, thought of as a specific and technological type of gaze. It is also important to look at the perceptions of those who are the object of the photograph. This is what I have done by taking the prohibitions concerning photography as the starting point for a reflection on the gaze.

By opening the tour with a statement about photographic instructions, the guides do not only remind tourists of basic ethical rules, but reaffirm, intentionally or not, their authority over the gaze. Observing how the photographs are taken makes it possible to realize this power. We have seen it in the case of the prohibitions, in its negative aspect therefore, but there is also a positive dimension of the instructions given by the guides. As for the prohibitions, the indications of the guide can be explicit or implicit. It is the guide who draws attention to this or that element of the scenery, and who can sometimes authorize a person to be photographed. This last case is, especially if we place it in the multitude of photos taken by each tourist, extremely rare. Most of the time, the subject of the photographs is only the scenery pointed out by the guide, such as the tangle of electric cables that has become a visual characteristic of the tourist favelas. It is assumed that here, the absence or rarity of shots of individuals is not explained by a lack of interest in this kind of photos, but by the nature of the situation in which they must be taken. The possibilities of photographing individuals are extremely limited precisely because the possibilities of the gaze itself are, in this situation, limited, if not completely non-existent.

While it is possible to link photography and the gaze in this way, it is because they have a set of properties in common. Of course, photography has something particular. Even by minimizing, as I did, the importance of photography becoming an object, this has nevertheless an undeniable impact. The photograph exerts a specific violence which exceeds in certain places that of the gaze. However, we can put forward the hypothesis that these properties have been transmitted to the gaze, but to the gaze of the tourists, not to any form of gaze. The tourist gaze, as a phenomenon perceptible to others, is never far from a photograph. It too, given the matrix of the tourist situation and its inequality in the tendencies of the actors (daily/non-daily, familiar/exotic, mobile/immobile), carries a value of danger and predation. For Susan Sontag, "every photograph is a *memento mori*" (Sontag 2008). According to Sontag, taking a photograph is to participate in making the mortal, changing and vulnerable character of another person more apparent; it is to bear witness to the implacable march of time. The tourist gaze equally has this harmful power, whether it inherits from photography or from the real and perceived association between photography and the tourist gaze – in which the photographic gaze represents a major part of what Urry named the tourist gaze, and which is more often simply the photographic gaze in tourist situations than a real innovation. Where the tourist gaze differs from the photographic gaze is precisely in the way it takes on the same attributes, culturally constructed in the practice of photography, without the need for a camera to be actually present. The dangers of the tourist gaze, its objectifying powers (thus pointing to impermanence), are more or less the same as those of photography. It is in this that we can say that the eyes of tourists are always cameras.

8.2.3. *The gazed and the gazers*

Around photography and the rules that are stated in order to control its impact in the tourist situations described here, I have gathered ethnographical elements that allow us to speak of the gaze, provided that we admit that the rules relative to photography ultimately apply to it. It is now appropriate to tackle the much more complex task of observing what can be said more directly about the gaze, without evoking the act of photography which, by materializing the gaze, makes it visible to others and to the ethnographer. In doing so, it is also a question of leaving the realm of discourse and basing interpretations on direct observation of tourists, through the intermediary of field notes that report a set of information relating to gazes, that is, *who* is looking at *whom* or *what*, and above all *how*: what are the modalities of the gaze, and are they modified according to the situation?

It is a question here of analyzing how gazes differ according to the spaces in which the tourists find themselves. The favela tours take various forms according to the guides, but in order to describe this phenomenon, I will concentrate on particular forms of excursions that take place mainly by minibus or car, with rare pedestrian outings in places considered important by the guides (visiting of stalls, crossing of alleys, access to overhanging viewpoints, etc.). Observing this type of tour, which is the most common form, allows us to consider two spaces in which the gaze is deployed in a different way: the minibus and the street, that is, as we shall see, a space of freedom and a space of constraint.

Because the tourists are in a situation of not being gazed at, the minibus is undeniably a space in which gazes are free. Here, the tourist looks at the inhabitants of the favela, who are gazed upon while not being aware of it. As a result, the tourists are entirely immersed in a position of spectatorship, which makes their gaze completely harmless (despite the fact that it is, from the omniscient point of view often adopted by critics, considered more intrusive, precisely because it is hidden, and it is partly here that the impression of voyeurism is born). Inside the minibus, the eyes are mostly turned towards the outside. The tourists wriggle in their seats to try to see the inside of a bar, or the passage of an alley. They gaze, with a wandering eye, at a game of cards on the side of the road. Attempts are made to take photos through the window, without having to ask permission. These behaviors have everything to do with the positions that the different actors occupy within the physical space and the sharing of the roles of being gazed and gazers, and with the conscience that the individuals have of their position in this visual social space. In this regard, it is interesting to observe how the situation differs in the case of another type of tour offered by the company Jeep Adventure. Because they cross the favela in a Jeep, that is, by means of an open vehicle, the tourists seem to apply a more drastic constraint to their gaze. The following field note describes what changes in this case:

> Sitting in the Jeep, tourists often seem to feel a certain discomfort and seem to manage to avoid meeting the eyes of the inhabitants they pass on the roadside. The way they sit, on seats parallel to the road, has a lot to do with it. But it is obvious that when we have not yet arrived in the favela or when we cross more deserted corners, the tourists contort their bodies, taking pictures. But when we arrive in places where there are a lot of people, the tourists tend to immediately redirect their gaze to the inside of the Jeep, resuming their discussions or looking intently at the guide, who has either taken a seat or is turning to the front and back again (Field note, Rocinha, August 2016).

The situation in the minibus, a closed vehicle whose presence is common throughout the city, and to which the inhabitants do not pay much attention, is very different. The eyes, assured of not crossing others, are freer, and also more insistent. The guide, in this space, does not manage to capture a visual attention and becomes a kind of voice-over on which the tourists contemplate the outside, scrutinizing so much in detail the scenes of the daily life of the favela, often without a direct link with the guide's discourse.

On the other hand, the situation of crossing a busy street is very different, since everyone is both looking and gazing. Observing the gazes when tourists are in the favela, in the middle of a public space dense in population, suggests a certain uneasiness. At various points in the favela, tourists are invited to get out of the minibus with the guide. The purpose of these stops is to allow the guide to explain the functioning of the favela around a variety of themes (electricity, commercial life, garbage management, etc.), and intentionally to counter the sensation that tourists may have experienced while crossing the favela in a bubble. When the minibus stops, the guide usually gets out first, and then stands five or six meters ahead, on the side of the road or at the intersection of an alley. The group follows, and the tourists get out and walk along the bus in a line until they find the guide. There, the group positions itself, huddled together – it is rare for a tourist to venture away from the group – and the visitors even hurry to take their position in a circle around the guide. What happens in terms of gazes follows entirely the logic of the physical disposition that the group spontaneously takes. It is flagrant, indeed, that the guide captures most of the gazes. The tourist gaze remains riveted on the guide and escapes from the limits of the circle formed by the group only to rest on the various points explicitly indicated by the guide. No gaze then crosses that of the inhabitants who, as far as they are concerned, do not deprive themselves to observe the group, sometimes visibly amused. It is noteworthy, moreover, that this situation is almost reversed when the group crosses an empty small alleyway. There, the inhabitants who cross the group rarely initiate a visual interaction and are content to continue on their way, looking at the horizon or at their feet. In the busy streets, it is the tourists who are careful not to look at others. The much more sparing way in which they photograph here corroborates this description of gazes: the cameras carefully avoid the inhabitants and only appear when the guide points out something in the favela that can be easily photographed, without anyone having the impression of being the real object. Here again, therefore, in accordance with the idea developed earlier, what applies to photography seems to apply to the gaze in general.

What changes, between the different spaces, is the awareness of the different actors in the game and their positions of gazing or being gazed at. These positions have three combinations: gazing while being gazed at, gazing while not being gazed at and not gazing while being gazed at. If the gaze of the tourists, in the public space, focuses so much on the guide, it is not so much to show respect for their discourse and their instructions as to show all the potential observers that the excursion is taking place in accordance with ethical rules. Also, paradoxically, we could say that it is the way in which a tourist is gazed at (in the situation of being gazed at, it is enough to be potentially gazed at) that conditions the way in which they gaze (and not therefore the object of their gaze). The situation, here, obviously concerns two fundamental elements of the tourist practices in the favelas, or in other words with the two major concerns encountered by the tourists: ethical and security concerns. Thus, by only looking at the guide or what he points to, tourists simultaneously perform two tasks. First, they confirm the authority of the guide and demonstrate their control over what is and is not made visible during the tour. Second, they make their membership in the group more obvious. This group, made obvious not only by the appearance of the tourists but also by their visible differences from the inhabitants, is here understood and made intelligible to others as the gathering of people who gaze at the same things.

The difference in the level of the insistence and the freedom of the gaze according to the two spaces is flagrant. In the second situation, the gaze is restrained; it is captured by an individual, directed by their instructions and is thus transformed into applied gaze, docile, in conformity with the standard. In the first, that of the minibus, it is contemplative, strolling, free, individual and independent of the instructions and the discourse of the guide. What conditions the difference of the gaze in these two situations is the difference of sharing between the roles of gazers and the gazed, and more still the difference of consciousness of this role on several levels. In this perspective, it is possible to link our observations with certain works of the current of symbolic interactionism, and more precisely with Goffmanian microsociology. Indeed, Goffman (1979) has analyzed the way in which individuals act when they are in the position of being gazed at, showing in particular the efforts they make to "keep face" when they are aware of their presence in the field of vision of others. In this sense, the position of being gazed at, as it appears in particular in our ethnography, when the tourists focus their gaze exclusively on the guide in the public space, also has to do with what the sociologists of ethnomethodology have called accountability to point out the ways in which individuals make their actions intelligible to others (Garfinkel 1967). In another way, Goffman shows the role that the gaze plays in social interactions: for individuals, it is a question, depending on the situation, of asserting their gaze (to attest, for example, that they are attentive to the discourse of the other or to the simple presence of the other), or, on the contrary, of diverting it so as not to impose the constraint of the gaze on the other and so as to introduce what Goffman calls "civil inattention".

8.3. The space of the gaze

Far from being entirely free, the deployment of the gaze is thus conditioned by a set of rules that emanate directly from the interaction and the relationship to the space.

8.3.1. *The modalities of the gaze*

Ethnography generally gives the gaze a marginal place, and this for easily understandable reasons, the main one being the difficulty of describing a gaze, which is summarized by Michel de Certeau's idea of the impossibility of a non-narrative nomination of the gaze. It is indeed difficult to envisage an ethnography of the gaze, and this difficulty is often concretized by a tendency to rise to such abstract generalities that the description of the gaze or gazes is exhausted in metaphors, as if there were nothing left in the observable situations of the gazes that would create a phenomenon. In other words, it is a question of momentarily dissecting all the actors/representations/thinking/gazes in order to retain only the gaze as an element that possesses a meaning in the social situation. By attempting to describe these gazes, and especially the way in which they reveal variations relative to the modifications of the context, it is a question of isolating which, in the gaze, both reveals and constructs the situation of interaction.

The chapter that Sartre devotes to the gaze in *Being and Nothingness* allows us to explain in what perspective the gaze can be used in ethnology. In this perspective, it is indeed a question of using the singular: the gaze rather than the gazes. For Sartre, the gaze does not matter in its embodied dimension and, moreover, he repeats several times that "it is never eyes which look at us" (Sartre 1984, p. 236) but always a "prenumerical" other (ibid., p. 239), in the sense that what counts is being aware of being potentially looked at, independently of the number of people who are really looking. He shows, moreover, that the actual presence of another person is not necessary for us to feel *in the presence* of the gaze: if I think I feel a presence behind me, which surprises me while I am spying through a keyhole, I do indeed feel a sense of fear or shame, even though the impression that someone is looking at me turns out to be false. The gaze thus ends up taking on for Sartre, in public situations, the appearance "of an intangible reality, fleeting and omnipresent" (ibid., p. 239).

One way of opening up ethnography to the consideration of the gaze is therefore to apply this point of view (which makes a real gaze equivalent to a potential gaze), which, at least as far as the field presented is concerned, is to be put in relation to the danger that the gaze implies, as the previous descriptions have shown. The ethnography of the gaze must therefore pass through the study of the positions occupied by the individuals in the visible space – the space of the gaze – which conditions the possibilities and ways of gazing. What such a perspective brings to light is that if the gaze is free in principle, it depends more on the structure of the space and the interactions than on its potential objects. Indeed, I have shown that photography could be considered as a particularly sensitive form of the gaze, and therefore subject to rules that can be explicitly stated, often in the form of an appeal to “common sense”. In the same way, I have described how the gaze unfolds more freely (and also, for example, with greater insistence) in individuals who are not in a situation where they are themselves being gazed at, that is, not in a position that requires the accountability of their gaze. Conversely, the position of being gazed at can, depending on the situation, lead to an inhibition of the gaze, which can be a complete inhibition. These situations are present everywhere in the social world, but are manifested differently according to a multitude of factors, from the strictly material nature of the space to a set of elements rather pertaining to the culture which fixes the symbolic outlines of the interpretation of the gazes.

So far, I have put aside the question of the cultural and symbolic determinations of the gaze. In the same way, I have not considered the gaze as it links a subject to an object, but as a phenomenon which appears in situations of interactions, according to varied forms and relative to the modifications of the general context of this appearance. Finally, and this rather in connection with the inductive method of ethnography than with the principles of phenomenology, to observe and describe the gazes in the favelas, drawing some lessons, of which there is no reason to doubt for the continuation:

– The gaze carries a danger indicated by the way in which the guides frame it (by the intermediary of the photograph) with a set of rules. This danger weighs simultaneously on the transmitter and the recipient of the gaze.

– Three positions relative to the gaze exist for individuals who enter the social space. These positions are: gazing at the gazed, gazing at the non-gazed and not gazing at the gazed. In a sense, these positions even condition the entry into the social space, since the last combination (not looking at looking) is not an interactional posture.

– These positions change the way of being, so that an individual does not act in the same way according to whether he is conscious or not of being gazed at and that, by the same logic, he does not gaze in the same way according to whether he is gazed at or not. A gaze, in that, is a gesture whose deployment depends on situations.

– To have the feeling of being gazed at, a potentiality of a gaze is equivalent to an observed gaze.

These different components, linked together, make it possible to draw a set of conclusions concerning the relations between the gaze and the different cultural and interactional determinisms that weigh on the tourist encounter. Here, we assume that the gaze has a dual function. On the one hand, it reveals the structure of the tourist situation: its rules (even tacit rules, that is the point), its power relations, the way in which it invests in physical space and with which it is shared in it, etc. On the other hand, it reveals the way in which the tourist situation is structured, the way in which it is shared, etc. In that, the modalities of the exchange of the gazes are themselves produced by the nature of the tourist situation. On the other hand, the gazes constantly produce situations. Non-verbal forms of interaction, but no less essential, are contained in the exchange of gazes. This is an essential part of the interest of an ethnography of gazes: they are not only produced by the matrix of the situation, but produced in return situations which sometimes summarize themselves there.

The sum of the reported rules and the behaviors described in their variability confirms that the gaze exerts a symbolic violence. We can consider that the tourist situation is structured in part with the aim that such or such group of individuals escape this violence.

8.3.2. *The danger of the gaze*

Through the ethnography of tourist situations in the favelas of Rio de Janeiro, I tried to show that, taken as an object of observation, the gaze can be approached from the angle of the ways in which it is constrained. I first showed that the rules of conduct issued for taking photographs could be considered as being valid in general for gazing, which implies that photographing and gazing induce the same type of relationship with the other. In a book devoted to the affinities between power and the gaze, Gérard Leclerc writes that "photography is a bit of a technological incarnation of the predatory function of the gaze, at least during the act of taking a photo" (Leclerc 2006, p. 2). It is useful to keep in mind the idea of the "predatory function of the gaze". I then described how the gaze was dependent, in its deployment, on the specific situations in which the various actors were immersed. More specifically, I put forward the idea that the gaze depends on the understanding of the sharing of the roles of the gazer/gazed. From this point of view, the gaze appears as an element of social interactions which, like any other means of communication, requires the knowledge and application of rules of conduct. In this situation, the gaze is an important issue and all the behaviors observed lead us to consider that it involves a significant danger. Indeed, at least part of the ethical balance of the situation is built at the heart of strategies (tacit or explicit, self-stipulated or imposed) of neutralization of this danger which seems inherent to the gaze.

The idea that danger can emanate from the eyes is not new. It has been well represented in several myths since antiquity. In anthropology, the literature dealing with the notion of taboo has also multiplied the examples of prohibitions in the field of the gaze. In an article devoted to the gaze, Françoise Héritier points out the fact that men are free to gaze at both other men and women, whereas women almost nowhere enjoy this possibility (Héritier 2004). In many of these examples, the gaze is a major issue, not only for the order of society but also directly at the level of individuals, for whom it is often life-threatening. The ritual prohibitions relating to the gaze speak volumes about the risks associated with it, as in the case of the excursions commented on. In addition, it is interesting to note that the gaze is most often perceived as a danger in both directions, since it weighs on both the one who looks and what is looked at.

These two sides of the risk are present in the myths of Ancient Greece, where we find several occurrences of transgressions by the gaze. Among these myths, the most famous is that of Orpheus. Unable to comply with the conditions set to bring back his wife Eurydice from the Underworld, he kills with one glance the one for whom he had gone so far as to defy the gods. Here, the danger of the gaze weighs first on the object. If we compare the instructions given to Orpheus by the god of the Underworld to those given to tourists in the favelas concerning photography, it would seem that the danger that they seek to avoid hangs rather on the eventual object: the transgression of the rules involves a risk of destruction for the one looked at. The meaning of danger is reversed in the myth of Actaeon, which tells of a hunter who caught Artemis naked taking a bath. Here, divine vengeance is realized when Actaeon, transformed by Artemis into a stag, is devoured by his dogs. In this case therefore, the danger pointed out weighs well on the one who gazes – by extension, it is the tourists that the instructions protect and it is them that a transgression would put in danger. Similarly, in this sense, is the myth of Medusa, the Gorgon who petrifies anyone who crosses her gaze. The myth of Medusa is particularly interesting in the stratagem employed by Perseus to kill her: he tricks himself by looking at his reflection and not directly at the Gorgon, and thus introduces a fundamental difference between gazing and seeing. It is not seeing that is dangerous for Perseus, it is gazing. Thus, intentionality is of central importance, and we find it expressed in the behaviors observed in the favela. Nothing is, apparently, directly prohibited and excluded from the field of vision. So we can make the following hypothesis: if it is on photography that the only explicit prohibitions weigh, it is because photography induces not only a greater visibility of the gaze, but because it also implies in an undeniable way an intentionality.

In the ancient myths briefly mentioned, which remain popular legends, the two dimensions of the danger relating to the gaze are thus expressed: by gazing at what is prohibited, we risk at the same time destroying the object and destroying ourselves. The stakes are serious, so much so that to a mythological corpus renewed many times and to a quasi-universal conscience of these dangers, are added new warnings specific to the situations, as well as to the new technologies which modify the frameworks and the modalities of the gaze.

Concerning the two extremes of danger in the gazed/gazer relationship, I have already established that the danger for the beholder consists mainly of an exposure to the sanctions – if only to the disapproval – entailed by the transgression of the rules. But, and it is essential to understand this, what about the danger incurred by the people being gazed at? If there is something that can be considered as rules, what are they supposed to protect against? This raises, in the end, the following question: what kind of danger is induced by the gaze? There are many answers to this question. Sometimes the issue is directly the general social cohesion. But it is often the individual who is vulnerable. It seems, in fact, that the gaze can first of all harm the individual, which of course is at the heart of the logic of avoiding the gaze, which we all experience in our confrontations with strangers. Jean-Paul Sartre even sees in the gaze the origin of the first psychological and philosophical disturbances of the individual, the first denial of humanity. Indeed, it is the gaze of the other which makes of me a thing, insofar as, according to certain modalities, it transforms me into a thing that I am for another subject: the gaze of the other brings about the consubstantial report of my existence as the other. It is even in this objectification painfully undergone in the gaze of others that "hell is other people" (Sartre 1989) takes root. If we can question the universality of the feeling of dehumanization that Sartre depicts, it is allowed to see in its wake a set of inconveniences that the faculty of objectivation of the gaze of others causes. In our ethnography, the risk that the gaze involves seems to come partly from this dimension: the gaze transforms into an object and returns to its strict condition of the point of view of others. It is appropriate here to point out the existence of a criticism external to the tours (external in that it is, in these terms in any case, more often the fact of other travelers than of the inhabitants of the favelas), which points out the ethical problem that constitutes the transformation of disadvantaged populations into objects for tourism. The gaze could then be understood as the concrete phenomenon through which this transformation occurs. On the other hand, the objectifying powers of the gaze also threaten the tourist experience. According to Sartre, one of the inconveniences of the gaze of the other lies in the way it prevents us from mastering our perception of the world: "The Other's presence in his look-looking […] undoes the world by the very fact that it causes the world to escape me" (ibid., p. 272). Or again: "The alienation of myself, which is the act of *being-looked-at*, involves the alienation of the world which I organize" (ibid., p. 263).

8.3.3. *Space of the gaze and authenticity*

The observable norms of the gaze seem thus to be able to be defined as the result of the incorporation of the rules supposed to protect from the gaze in two senses: for the gazer, whose authorization to penetrate the exotic space depends on their faculty to respect them; for the gazed at whose dignity is protected by these rules. In the descriptions given so far, it is obvious that the positions of the gazed and the gazer after are interchangeable and that the inhabitants cannot be limited to the position of being gazed at. In tourist situations of this type, the gaze is effectively mutual (Maoz 2006). The fact that it is mutual, and that the gaze of the tourists is therefore itself subject to the gaze of the inhabitants, is even perceived as an ideal rule to ensure the good ethics of the tour. Indeed, it is striking to observe that the critics of tourism in the favelas point first and foremost to the motorized favela tours, because of the possibility they offer to tourists to extract themselves momentarily, inside the vehicles, from the gaze of the inhabitants, while continuing, on the contrary, to watch them. It is this fact, this non-reciprocity of the gaze, that best represents what the critics of the favela tours mean by voyeurism when they criticize them. The accusation of voyeurism rests therefore entirely on the modalities of the gaze and on the questioning of the existence of non-gazed gazers. On the other hand, it says nothing about the object of this "voyeurism", the same that is at work when tourists look inside the houses, and therefore about what the tourists can gain from this non-reciprocal gaze. It is again on the side of authenticity that we must look for the motivations of these transgressions.

In the context of tourism, the fact of being in the position of both the gazer and the gazed at can therefore lead to a contradiction. This is particularly true in the context of the construction of an authenticity of the object of the tourist visit. The experience of authenticity functions in a configuration that clearly posits the gazers (who could be called spectators) and the gazed at (the object of the spectacle, therefore). But when the positions can be reversed, authenticity is threatened. This is why it seems to me right to consider that the nature of this situation (constituted by the construction of an authentic otherness) entails in some way the setting up of a logic of avoidance. It is moreover the sense that we can interpret from what Georg Simmel affirms when he writes:

> From men whom we merely see, we construct a general notion with infinitely greater ease than if we were to speak to each of them (Simmel 1999, pp. 636–637).

Here, Simmel evokes above all a difference between visual and verbal interactions, but he clearly shows the objectifying power of the gaze. In the case of tourism, this power is both a threat (that of seeing ourselves transformed into an object) and a means of satisfying the construction of a narrative and of perceiving the other in a perspective that conforms to it. In other words, the maintenance of certain symbolic narratives, which form the framework of favela tourism, depends in part on these moments when tourists can unilaterally deploy their gaze. When asked afterwards about the highlights of their visit, tourists often mention these small scenes seen in the distance, without their presence interfering with the normal course of daily life. It is even common for tourists to prioritize a hierarchy of authenticity that favors these unmediated observations over elements of the guides' discourse:

> What the guidebook says is very interesting, but we could have it in another way. In my opinion, that's not why we have come. What I will remember are small moments that made me experience life in the favelas. Children playing soccer, for example. Also, at one point in an alley, we passed a house and inside there were two teenagers dancing. These are the little moments that, for me, represent the essence of the favela, and you can't understand this if you stay outside.

Another such example:

> I'm not entirely convinced by what the guidebook says. It always says: "The state does this, the state does that", even when talking about banks and telephony – and it never talks about the private companies. In short, it's just that the guide is absorbed in his idea that the state is responsible for everything. And OK, that's true for the most part, but it leaves a lot out. I feel like he wants to fit everything into the story he's telling us. Actually, that's not the most interesting thing. For example, he tells us that there is a graffiti artist in the favela; it would be nice to be able to visit the workshop. On the one hand, the guide is useful for getting around, but at the same time, it makes us miss a lot of things that we [only] get a glimpse of by looking in from the side.

These types of comments which question the veracity of the guides' discourses are very frequent. And it often happens that elements that are fleetingly glimpsed and on which the guide does not dwell are contrasted. Most of these elements come from moments when the tourist's gaze does not follow the guide's discourse, where it ventures freely, and in particular in a configuration where the tourist's gaze is not subject to the gaze of others, as is the case in the first example, which refers to a scene witnessed inside a house. This suggests a link between the transgression of the gaze, its non-reciprocity and the feeling of authenticity. In this, the examination of the gaze says something about where tourists a priori locate authenticity. It does so in a way that is different from and complementary to the way Dean MacCannell drew it.

To define authenticity, MacCannell borrows from Erving Goffman the idea of a separation between stage and backstage. For Goffman (1979), the stage (front) is the public place in which individuals play a role under the gaze of others. Knowing that they are potentially being observed, they apply themselves with more or less mastery to the role assigned to them. The backstage is the private and intimate place where individuals withdraw. Since they no longer have a role to play, they can rest there. The division between stage and backstage is social, but it can take physical forms. Thus, the domestic space of the house, closed off to eyes, is the typical example of a backstage. For MacCannell, authenticity is always thought of as a natural holding to what takes place behind the scenes. According to him, our societies still think of the real and the authentic as being situated on the side of the intimate. The feeling of authenticity therefore always emanates from a dialectic between stage and backstage. MacCannell has developed a typology (MacCannell 1973) that allows for the insertion of different layers between the stage and the backstage, because, as a result of the tourist search for the backstage, the individuals who are the objects of the tourist gaze construct intermediary scenes so that they resemble the backstage. It is this phenomenon that MacCannell calls staged authenticity. From the same borrowing of the Goffmanian dichotomy, the data presented on gazes can lead in another direction.

Indeed, in order to understand the emergence of the feeling of authenticity in the favela tours, I propose to look for a way to complete the dichotomy of authenticity/staged authenticity (MacCannell's). The favela tours, as I have already shown, are best defined as reality tours, whose object is a hidden reality, unknown and not altered by its setting in tourism. Of these three qualities, the last one appears as the most important. The reality of the favelas, in fact, will be successively revealed and understood – that is the object of the tour – but in no case should it be altered by the very enterprise of its discovery. The same is true of reality, of the normal course of things, as of otherness. The feeling of authenticity comes first of all from the assurance of not modifying reality by its presence. In this respect, the non-reciprocal gaze is conducive to the feeling of authenticity. Indeed, many discourses collected from tourists point out the importance of these gazes thrown towards scenes of the everyday life that the gaze did not alter, because it remained dissimulated and not anticipated. It is obvious that, from the point of view of Goffmanian dichotomy, what the tourists observe most often belongs to the stage and not to the backstage. The inhabitants, when they are on the street, are consciously in a public situation. Yet the stage, from the point of view of the tourists, and on the condition of a non-reciprocal gaze, has a status close to that of the backstage, of unstaged reality. Of course, the interior of the houses, which we have seen is sometimes discreetly observed by tourists, remains at a higher level of authenticity because it is a space of intimacy. To be able to look at it, whereas certain tours envisage entering a house, that of the guide, for example, is the source of a feeling of superior authenticity, precisely because the space is excluded a priori from the tour, and thus from the normalized space of the tourist gaze. In the context of tourism, we can therefore say that authenticity is reality minus the self.

9

Reality and the Tourism Frame

In Chapter 5, devoted to authenticity, I began by mentioning the existence of a broad lexical field that allows the notion of authenticity to emerge even when it is not explicitly present: real, true, profound, etc. Among tourists, these terms are all linked to the same evaluation of the tourist object. It is always a question of separating the real from the fake, the original from the counterfeit, the superficial from the profound, etc. These terms, in this respect, can appear to be, if not equivalents, at least linked to each other. From the point of view of tourists, it is not certain that authenticity and reality are notions covering very different meanings. Rather, the notions appear complementary: authenticity is the reality that we are sure of, not the way it is staged. This relationship between reality and authenticity appears to be an essential topic if we are to understand the tourist experience.

From a tourist's perspective, authenticity and reality seem inseparable – not equivalent, but complementary. In fact, this is one of the essential characteristics of favela tourism, whereas the correspondence between reality and authenticity may not be systematic in other practices, as some researchers claim (Brown 1999). In the case of favela tours, authenticity cannot be anything other than the experience of reality. This is undoubtedly why the term "reality tourism" is systematically linked to favela tours via its main promoters, since it is reality that is targeted. The articulated use of the two notions, reality and authenticity, is important for understanding how the tourist experience is formed. Authenticity and reality, although sometimes confused, nevertheless refer to two distinct ontologies – both from the tourist's point of view and from their own understanding of their experiences. First, from the tourist's point of view, we can consider that reality refers to something objective, independent from their point of view and pre-existing a personal experience, whereas authenticity comes about to qualify a part of this reality; subjective but attested as being part of this same independent whole. This is, more or less, the tourist's point of view on the relationship between reality and authenticity.

To study this, we can adopt a different posture. Reality thus appears as an inherently incoherent, heterogeneous whole whose existence is not given to consciousness. Authenticity concerns the way in which what is conceived as reality reaches the individual's consciousness. In other words, authenticity is one of the possible methods for understanding reality, both notions being culturally forged, that is, dependent on definitions subject to variation. Authenticity, in these terms, consists of the ordered result of the experience of reality, as well as the individual's consciousness, which makes it coherent, and divides up units of meaning in a heterogeneous flow of events, people, actions and places.

What is interesting to study in the social sciences is the collective structuring of this experience of reality as it emerges as authenticity. It can, of course, have individual foundations that are less directly linked to anthropology. To interpret the collective structuring of this experience, I put forward the notion of a frame, which intervenes at several levels. Indeed, the tourist experience of the favela only makes sense through the sharing of a collective frame – which is also (i) a narrative frame (the place of the favela in Western cosmology is partly given a priori), (ii) an interpretative frame (which conditions the way in which events reach consciousness during the experience) and (iii) an interactional frame (which dictates the behaviors to be adopted during the interaction).

9.1. Narrative frame and experience

In order to understand how the favela tours are structured, it is necessary to place the tourist experience in a "hermeneutic circle". In a work devoted to the anthropology of experience, Victor Turner and Edward Bruner define the hermeneutic circle by distinguishing three distinct orders: reality, experience and expression. Inspired by Dilthey, for whom "reality only exists for us in the facts of consciousness given by inner experience" (Dilthey 1976, p. 161), the two anthropologists distinguish between what is really there (reality), the way in which this reality presents itself to our consciousness (experience) and the way in which experience is framed and articulated with other experiences (expression). In other words, the hermeneutic circle has three dimensions: life as lived, life as experienced and life as told. Obviously, the relationship between these three dimensions of life is problematic, and it is this problem that constitutes for Dilthey the very object of social science. Our understanding of tourism practices should also be rooted in this hermeneutic circle. However, in the case of tourism, experience and narrative are intertwined in a particular way. To observe tourists' visits to favelas is to observe a transition between a reality (which cannot be postulated to have an essence) and the tourist experience that gives it its meaning. But this meaning is always conditioned by the narrative that, more than in any other social situation, takes center stage.

First, the narrative precedes the experience. This narrative is structured due to the way the favela is presented to tourists and because they are put into a market of otherness. Favela tourism is based on reality tourism and the valorization of an authentic otherness. This general narrative frame influences the structuring of practices and the frame within which tourists experience a reality that goes far beyond it, and that could be presented differently. However, tourists are often willing to find in the favela something other than what they expect to find there; indeed it is not the tourists themselves who (directly and exclusively) assign a coherent frame to a reality, which is actually heterogeneous. Rather, this frame is co-constructed, and it is primarily through the guide (i.e. the intermediary) that the narrative imprints its influence on the practices. Erik Cohen distinguishes five functions of the tour guide (Cohen 1985). The first, which is said to be instrumental, resides in the guide's power as they allow access to new areas and have the ability to direct tourists there. The social function aims to maintain the conviviality of the group, while the interactional function designates the guide's control over interactions between the locals and the tourists. Finally, the guide is able to point out places of interest and, above all, can interpret the sites according to tourist expectations. For Cohen, it is this last factor that is the most important, and the observation of excursions in the favelas corroborates this idea. The five functions designated by Cohen are all part of the essential communication skills for being a guide to a favela. But it is around the means of ensuring the communicative function that the selection seems to be made – if they want to succeed as a legitimate mediator between tourists and favelas, it is important the guide comes from an international background. Therefore, when it comes to providing an experience adapted to the grand narrative that underpins tourist attractiveness, the tour guide must have the ability to understand, as best as possible, the ways in which a favela can be interesting for Western tourists.

This does not mean, however, that the narrative will remain inexorably unchanged by the tourists' experience of the favela. On the one hand, the guides have a capacity for innovation that allows them to change some elements and add new ones, which will complete the narrative or modify it. On the other hand, once the excursion is over, the continuity of the narrative changes hands and goes to the tourists. This is the second key point of influence regarding the narrative, the one made by the tourists after their visit has ended. Already influenced by the narrative, the experience is again put into a narrative. I have already mentioned a few examples that show that small stories can help the big story evolve – notably by being told distinctively and thanks to the rhetoric of authenticity. These factors mean that the stories demand to be told differently, person-to-person, as we make new aspects known, etc. The experience of visiting a favela can be solicited, and thus placed in a narrative, to answer different stakes. In the example cited in Chapter 5, two forms of experiencing a favela are set against each other in a discussion that compares the relative values of the different participants as travelers. In this frame, we have seen that the content of the experience undergoes a modification in its narration: the concealment of the experience's touristic frame. I explained this omission to be a matter of distinction, and thus an awareness of the negative perception of so-called "tourist" activities. Nevertheless, it seems excessive to consider this concealment of the frame to be conscious manipulation. In fact, it is clear that this occurs as a modification of the experience by means of a narrative that is not only expressed, but that comes to form the very memory of the experience. The accompaniment of a guide and the presence of the group are concealed within the very consciousness of the visitor. This is not amnesia or repression, but simply the effect of a change of setting. Many of the elements that were necessary for the experience to take place, in particular those of a practical nature, no longer have a function when placed in a narrative, as the narrative usually focuses on the tour itself and the tourist's discovery (impressions, feelings, etc.).

This narrative phase is also essential because it contributes to the construction of the grand narrative, and thus to conditioning other experiences. In addition, it has a specific social dimension (as a moment valorizing the experience and as a moment of symbolic gain) that moves the journey forward, and also generates a particular type of narrative linked to the symbolic functions of the tour. Regarding this, it is indeed a hermeneutic circle in which reality, experience and narrative are always entangled, and not a cycle that would have the narrative as a final stage. When it comes to the way in which the favela appears to the tourists during their visits, this great narrative is fundamental and, regarding different aspects, it structures the frame of the tourist experience. Not only does this frame select some elements over others, because they support the narrative, but it also excludes some. Without these framings, the favela is of course a place like any other, that is, highly polysemic, heterogeneous, made of disparate elements and containing contradictory values. Any reification of the favela is a social operation, relative to a given group and to the symbolic rhetoric that presides over its way of constructing a (simple) meaning from a reality that is never totally graspable. In other words, the favela is never – and for no one – equivalent to its position on the map of its city, nor to the sum of its streets, houses and inhabitants, which forms an ontological reality in the sense of what is there independent of the embodied points of view. What it is depends on the frame constructed for specific experiences. As such, the tourist experience frame and that of its narrative form two different social situations. The fact that, in spite of some important modifications that take place over the course of the transition from one to the other, the so-called grand narrative continues to hold sway and does not attest to its veracity. It simply shows its function, which is to unify experience and narrative by giving the impression that they correspond to the same reality – which is apparently homogeneous and coherent. If we consider experience and narrative to be two complementary phases, it is certain that they maintain a different relationship with reality. While the relationship is more free during the narrative, the sense of coherence of reality is more threatened at the moment of experience since, as

Husserl wrote, "what is currently perceived is surrounded by an obscurely conscious horizon of indeterminate reality" (Husserl 2018, p. 89). In order to understand how the tourist experience of a favela can hold onto the grand narrative in the face of the inevitable inconsistencies of reality, it is necessary to consider the nature of the frame of this experience.

9.2. The tourism experience frame

What about the nature of the tourist frame? At first glance, it refers to a situation that Erving Goffman commented on abundantly: that of the theater. In the case of theater, a difference is marked between spectators and performers. A line of separation is maintained which, most of the time, takes a physical form. Spectators are not allowed to intervene in the performance and their various authorized manifestations, applause, for example, can be easily ignored from the stage, where the performers do not interrupt the continuation of their performance. This purposeful lack of attention directed to off-stage noise depends on the maintenance of the theatrical frame a responsibility that falls mostly to the performers. The tourist frame does not function in the same way, mainly because of the role played by authenticity or, to put it more in the language of frame analysis, reality. Indeed, the different actors involved in the theatrical setting agree to evolve in a transformed setting[1]: what is on stage replicates reality, but is not reality (Goffman 1974).

No such agreement structures the tourist experience of the favelas. The favelas' inhabitants are not performers, neither from their point of view nor from that of the tourists. They have to follow the course of their normal lives, because it is precisely this normality that is sought after by the tourists. The tourist, on the other hand, takes the form of a spectator whose presence must not influence the course of the activities on stage. While the inhabitant is not fully, in the literal sense, a performer, this is not so much because of the lack of division between the scene and the spectators, but more so because the inhabitant would not be able to control the operations of framing and does not have the skills to do this. This responsibility falls instead mainly on the guide, who makes the situation noteworthy. Moreover, the entire role of the guide could well be summarized by their actions relating to the establishment of the tourist frame, that is, a frame whose aim is to link values to observable facts, transformed thus into symbols. Except, as previously mentioned,

1 In *Frame Analysis*, Goffman contrasts primary and transformed frames. Primary frames are "interpretive schemas" acquired by each individual, and which allow him to interpret social situations. Transformed frames are modifications of a primary frame (a sentence with a double meaning, a dress rehearsal, etc.).

the narrative precedes the experience, and the symbolic setting of the favela is already achieved in a general way.

9.2.1. *The guide, a professional when it comes to framing*

The symbolic foundations of the favela tours, analyzed in the previous chapters, precede and structure the excursions. Although visiting a favela, as a local guide in Rocinha put it, consists of "seeing reality as it really is", the favela tours present only a partial vision of reality, putting forward what best corresponds to the narrative frame given by the symbolic place, all taken by the favela in an economy of otherness. The idea of reality, crystallized by the term "reality tour", is crucial for favela tours, and it is out of the question for guides to stage things in order to support the narrative (I have shown some examples of this, such as a spontaneity of a culture of poverty, or signs of the real "Brazilianness" of the favela). Rather, the guides must deal with the vagaries of the course of daily life in the favelas and know how to place them within the frame of the tourist narrative, even though contradictory events sometimes occur.

Favela tours are frequently criticized for not generating any real contact between the tourists and the favela's inhabitants. The excursions are criticized for their superficiality, all the more so if motorized transport is used. The encounter, indeed, does not seem to be part of the favela tour program, which is mainly organized around the guide – the guide is the only intermediary between the favela and their visitors. They present themselves as essential for crossing the borders between the "normal" world and that of the favela, between the familiar and the unknown, and in a way between the layperson and the sacred. It is a role that they assume from the beginning of the visits, taking advantage of a whole lexicon of exploration that gets used in brochures and descriptions.

The guides are also indispensable in the face of the need to reassure tourists. They help calm the frequent security concerns linked to the reputation of favelas as dangerous places. This reputation is only moderately shaken up by the guides, and sometimes they even skillfully reinforce it, no doubt for fear of seeing independent tourist practices becoming established in the favelas. At the same time, the guide assumes a position of protector that they quickly add to their role as the holder of knowledge. The opposition between the favela and the tourist world makes the presence of a guide essential, as a sort of interpreter between the two cultures. This skill, although more common, is often considered to be more important than protecting the tourists. Thus, when a tourist asks the guide, which happens relatively frequently, if it would have been possible to visit the favela alone, and the guide wishes to avoid the subject of violence, the latter will often reply: "Nothing would have happened to you, but you wouldn't have had all this information and, worse, you would surely have gotten lost." The guide thus appears as the one who will allow tourism to generate meaning from elements that, without the tour guide, would have remained too homogeneous.

Nevertheless, even though a large part of the guides' discourse exaggerates and romanticizes the difference between the tourist's world and that of the favela for strategic purposes, it is also true that not everything they say is invented and that very real obstacles hinder the possibility of an encounter between tourists and inhabitants. Thus, for tourists, inhabitants and guides, it is certain that only the latter possesses the intercultural skills that enable them to transition from one world to the other. Among the many skills and qualities inherent in any tour guide, multilingualism is obviously paramount. On the other hand, it is about as rare to find Portuguese speakers among the tourists as it is to find English-speaking favela inhabitants (and even then, the tourists have to be English speakers as well). In any case, meeting with the inhabitants, even though only through participating in daily scenes of favela life, is a rare and underdeveloped element. It is likely that this aspect of the favela tours generates a large part of the criticism leveled at them, which denounces the passivity of the participants and discredits the value of the experience. It is also clear that this very absence of encounter is a key aspect in the development of alternative practices (built on the favela tours' different strategies, which represent the initial form of tourist consumption for the favela) that travelers engage in, which differs in many ways from the strategies of guided tours, but ultimately little in terms of what attracts tourists to the favela. However, there is sometimes an absence of meaningful encounters and a lack of locals' participation, which in many cases is not even embodied by the guide (if they are not a resident themselves). In cases like this, interaction is limited to the souvenir shop built for tourism, though this has rarely been mentioned by the participants. Only in exceptional circumstances did the people I spoke to openly deplore the superficiality of the visit.

The favela tour frame does not require active participation of the inhabitants and, in many aspects, this absence of participation even appears to be a positive element that contributes to the tourists' impression of authenticity. As a general rule, the favela tour is less like an encounter between tourists and inhabitants than it is a place for tourists to interact with situational discourse. The favela and its inhabitants are a backdrop for a discussion led by the guide, whose aim is to associate the favela with values that make the tourist experience a valued one in a valued space. Outside of this, the main function of the guide seems to be to ensure coherence between the discourse and the course of the tour. In Goffmanian terms, the guide is the actor responsible for framings. They are in charge of defusing or counterbalancing what inevitably comes up against the integrity of the frame constructed by the tour guide. Most of the time, the frame is easily maintained and does not appear fake to tourists. A significant part of what maintains this comes from a relationship of avoidance that tourists and inhabitants mutually practice, allowing the guide to assume the role of a central figure rather than that of a simple intermediary figure. This is because, it must be emphasized, the guide has imperatives on both sides of the relationship and must demonstrate their ability to frame both the local inhabitants, who expect integrity (in terms of what is said, if they can access it from various clues, as well as in terms of the degree of intrusion represented by the visits of a given guide), and the tourists, who hope to experience a situation that is as authentic as possible; authentic being understood here as the least altered by their presence. The respective expectations of locals and tourists towards the guide give them this filtering power and, beyond that, explain why avoidance becomes the norm; each potential interaction between tourists and locals having the power to destroy the tourist experience frame. Most of the time, because they agree to leave the responsibility of framing to the guide, avoidance prevails and the guide is free to point out the aspects relevant to their tour's discourse or, in an opposite sense, they can illustrate their discourse by highlighting this or that scene of daily life that they come across. The guide appears here in their capacity of a specialist, whose partial importance Goffman points out: "In our society, we often give to the specialists the task to clarify an ambiguity of framing" (ibid., p. 303). Goffman gives the example of the forensic scientist, who makes it possible to place an individual's cause of death in the correct frame. During the favela tours, many events occur without the tourists being able to make sense of them at a glance. On the contrary, as a specialist who has been given full explanatory powers, the guide can explicitly provide tourists with the frame within which to interpret the event.

9.2.2. *Framing and reframing*

One of the fundamental points in the tourist narrative of the favela is the representation of happy inhabitants who are proud to live in the favela. This point is essential both to settle the ethical questions of tourists ("the inhabitants are welcoming and happy to show their home to tourists"), as well as to reverse the perception of favelas by transforming a stigma into pride. However, obviously, not all favela residents *are* satisfied with their reality. Most of the time, they play along with tourists or totally avoid them. But sometimes, out of a desire to assert their discontent or due to circumstantial reasons, some inhabitants express their discontent and refuse to help maintain the narrative of the favela for tourists. In cases like this, it is the guide's job to open a discussion about the event with their group. The guide will add their own spin on things, adding new elements to inspire a new interpretation of the event.

I observed, during a tour, one of those frame breaks that guides sometimes have to deal with. One afternoon, during a tour with a small group of five tourists, Zezinho, as usual, called out to three men sitting on the edge of a low wall. He visibly clashed with the refusal of one of the men to cooperate in the enchantment of the favela:

> *Zezinho* [addressing the group]: Hi, guys! You like Rocinha?
>
> [All three men, at this point, simply smile, clearly amused by the situation].
>
> *Zezinho* [insisting and taking one of the men to task]: Do you like the favela? Do you like Rocinha?
>
> *First man*: Yes, I like it very much.
>
> *Zezinho* [giving a bracelet to the first one and now addressing the man in the middle]: And you, do you like Rocinha?
>
> *The second*: Yes, I like it [Zezinho gives him a bracelet].
>
> *Zezinho* [addressing the last man]: And you too?
>
> *Third man* [smiling but looking away]: No, I don't like it.
>
> *Zezinho*: Don't you like Rocinha?
>
> *The third man* [still in the same posture]: No, it's shit here. Who would want to live here?
>
> *Zezinho* [visibly depressed by the answer]: Come on, the favela is beautiful, it's a great community. Don't you like it?
>
> *The third man* [still not looking at Zezinho]: No. (Conversation in Portuguese during an excursion, July 2012)

The first two men burst into laughter after this exchange. Their attitude may suggest that the third man's refusal to tell Zezinho what he expected corresponded to a desire, if not to tell the truth, then to tell him about what he really thought of the favela, to mock and not cooperate with him, thus expressing contempt for his job, or at least a lack of willingness to participate in it. After the exchange, Zezinho's reaction was to move away from the three men, but not before giving a bracelet to the third man, who accepted it. He then resumed the tour. Zezinho, turning to the group, decided to recontextualize the event for his tour group in order to give them a completely different perception of the conversation: "This one is a joker, he always makes jokes like that. But don't mind him, that's how it is here, people in the favela are funny." By modifying a posteriori the frame of his interaction, and by presenting it as being jokey, Zezinho managed to conceal the man's refusal of comply with his request to affirm having pride for the favela. Not speaking Portuguese and thus not being able to understand exactly what the discussion was about, the tourists had no reason to question the explanation given by their guide.

Events of this nature sometimes occur when tourists and inhabitants cross paths. It is in the alleys that the complicated crossing undoubtedly makes the tourists feel the heaviness of their presence in a more marked way. On irregular ground, between a tangle of houses and small passages where they sometimes hit their head, the visitors feel awkward, and like intruders. Each time the group enters an alley, the guide insists: "Be careful where you put your feet! Watch your head! And when you pass someone, step to the side!" The size of the group is often problematic, in that they block the flow of passersby. When residents walk down the same alleys, we cannot help but notice a big difference in pace between the often busy and hurried residents and the contemplative tourists, looking up, trying to find a piece of sky over the houses or photographing the maze of electric cables.

When a resident arrives in front of the tourists, that is, when they go up the alley, since it is always towards the bottom of the favela that the groups circulate, the guide, who generally leads the way, greets the resident and asks the visitors to move to the side. Most of the time, the resident passes the tourists in absolute silence. Occasionally, a tourist or a local will give a greeting, which is politely answered. In the hundreds of times I have observed this type of situation, I have noted with wonder how the locals, especially if they are alone, never look at the tourists when they pass them and continue on their way, staring straight ahead or watching their steps. The irregularity of the path probably does not fully explain this absence of gaze, and it is likely that we find ourselves in a situation that causes discomfort for the resident. It is thus a situation contrary to those observed in more open and frequented spaces, at the edge of large passageways or crowded pubs, where the avoidance of the gaze is clearly at the initiative of the tourist. In the alleys, outnumbered, the resident generally makes do with an evasive greeting.

When a resident arrives from the back, and therefore walks down the alley in the same direction as the tourists, overtaking the group is actually more problematic. Without addressing them or muttering a "*licencia*" ("sorry"), the resident tries to make their way between the tourists, who often do not notice their presence right away. In addition to a certain feeling of discomfort already described, these recurring situations can sometimes be the source of obvious irritation, as in the following case:

> One day, while I was in one of these alleys with a group of about ten people, larger than usual, an inhabitant carrying a pile of empty cages

> that were to be used to store chickens had apparently been trying to pass us for some time. Arriving in the middle of the group, the man jostled one of the tourists who had stopped abruptly to take a picture of the cables passing over our heads, causing the cages to fall with a crash that immediately alerted the guide at the front. The man, as a result of much frustration regarding his attempt to bypass the group, became angry. The reaction of the guide was immediately to try, while helping him pick up his load, to calm the man down in a friendly tone. Refusing as best he could the excuses and the help of the guide, he hastened to collect his cages as best he could and hurried back on the path that was now left free to him by the tourists who, on one side and the other of the alley, were passively watching the scene. As soon as the passerby had left our sight, the guide, visibly offended and embarrassed by the rejection, turned to the person responsible for the accident, who had simply murmured some confused "sorry", to reassure him: "Don't worry, [and turning to the group] you didn't notice? I think he was a little drunk. Well, we're moving on, but be careful to give the locals room" (Field note, Rocinha, July 2015).

The tourist frame, which is maintained by the guide, thus leaves little room for local people to express disagreement with the tourist activity. Dominated by a positive narrative of otherness and discovery, the tourist experience absorbs these dissents through the guide, who has complete control over the interpretation of events. If locals' annoyances or criticisms regarding tourism are expressed during the tours in this kind of way, it is primarily *because* the guides are keen to reassure tourists of the general acceptance of their presence. However, other aspects of favela life are also ignored, this time because of a disconnect felt about the symbolic narrative on which the tours are structured.

9.3. Avoidance and exclusion

I have shown that the favela experience frame is underpinned by a narrative frame, itself designed to enhance the favela according to its conformity to the positive imaginary that has been constituted about it. In this way, the tourist frame encompasses both that which ensures coherence and promotes a positive experience of the favela, as well as a bias that selects certain elements and excludes others. I have shown this in the case of events and interactions that cause a shift due to the residents, but which are recontextualized in the tourist frame due to the intervention of the guide.

With regard to the tourism frame, it is necessary to distinguish two aspects, which are complementary but of a different nature. According to Goffman's definition, an experience is both related to the meaning given to the experience (What is happening? Is what I see real?) and to the behavior to be adopted in a given situation. By emphasizing the importance of the search for authenticity in the tourist experience, I have shown that tourists adopt a behavior characterized by a certain avoidance. Since the tourist presence can be perceived as a pollution of otherness by the Same, it tends to be minimized. When tourists are in a visible position in a public space, they tend to avoid looking at the residents, whereas they look at them longer in the opposite situation, when the look cannot be mutual. In a way, it is by pretending that the inhabitants are not there that they manage to pretend that they themselves are not physically present in a situation whose authenticity they would risk compromising. In theory, the inhabitants are not required to respect the same distance. Yet they themselves practice a comparable avoidance, which is rarely broken, as in the cases described above. Avoidance appears to be a fundamental characteristic of the frame. Thus, it is the interaction that constitutes the most common frame division.

The many activist initiatives of the inhabitants are not part of the tour program. Instead, the guides tend to try to extricate themselves from the political authorities. This is a problem in many favelas. In Morro dos Prazeres, for example, independent guides complain that they cannot avoid paying a tax that is systematically demanded of them. The case of Santa Marta, where community tourism is firmly established, is an exception. The political message is much stronger there. Here, the office of the *associação* (association) is frequently visited, and the struggle of the inhabitants for access to the rights of citizenship and for a change in living conditions is explained through a political demand.

However, in most cases (including in forms of tourism that present themselves as being local), the narrative construction of favela tourism and the development of what I have referred to as "logics of avoidance" contribute to the image of a population that is much more passive than it really is. For them, tourism does not really open up a possibility for them to express their concerns, sufferings and demands. On the contrary, the rhetoric of authenticity and exoticism, because it is a projection of values, can sometimes contribute to further dismissing the inhabitants' resistance to their condition. Moreover, the valorization of marginality as a source of authenticity also involves working in subtle, if paradoxical, ways to maintain both symbolic and physical boundaries.

I said at the beginning of this book that, in the case of favelas, the spaces concerned by tourism do not enter the tourist economy according to their individual specificities, but due to the fact they belong to the favela category. This is one of the effects of the way in which the favela is promoted as a tourist attraction, and of the type of narrative and symbolization that constitute its structuring principles. It is, in fact, on the scale of the category that tourist symbolism has been constructed. One of the strengths of the narrative frame is therefore ensuring that a visit to a favela is an experience that is valid for the *entire* favela category. Some favelas, of course, rely on particularities and put them forward (Morro dos Prazeres and the street art, Santa Marta and the statue of Michael Jackson, Rocinha and the city within the city, etc.), but without ever losing sight of the category or the idea that tourists want to get to know the favela and will most likely only visit one. The Cariocas of the *asfalto*, some of whom are very critical of favela tours and tourists, and make a mockery of them, often hold this fact against tourists and assert a difference in nature between the real favelas and the "chic" favelas that tourists visit:

> *Me*: Why do you think tourists go there?
>
> *Orlando*: Because they are crazy. They don't know. They come to Rio, they are there for a week. I live here, I know that there is a big chance that something will happen to me if I go there. It is dangerous. Don't you feel it?
>
> *Me*: Of course there are dangers. But when I go there with groups, there are no problems, no more stories than elsewhere.
>
> *Orlando*: Pfft. Yes, but you are talking about Vidigal Dona Marta, that's favela chic (Discussion with Orlando, engineer and military personnel, July 2015).

Orlando, who questions the fact that tourists visit authentic favelas, does not perceive the favela to be exotic at all. This type of discourse is frequent, and it shows a recurrent tendency to separate the favelas visited within the category. Also, what is a conscious argument of the promoters of favela tours (the tour of a favela can improve the image of the entire category) seems to work only from the Western point of view, and does not leave the tourist frame outside of which the stigmatization persists. However, it must be noted that this all-encompassing symbolism promoted by tourism has a lot to do with the exclusion of the local residents from the tourist frame.

Indeed, while *favelado* identity can be claimed in some arenas, it is not predominant. Demands and political struggles most often involve denouncing local problems, which affect specifically a space, and not the whole category. In the name of the representativeness of the favela visited at the scale of the category, it is often these fights that are silenced, and the actors who carry them are made invisible. Through the praise of the adaptation of inhabitants, the valorization of the culture of poverty hides the active resistances in favor of a philosophy of acceptance. And the least we can say is that a fine line separates acceptance and resignation in the face of poverty. The guides' way of treating the issue of drug trafficking is quite significant.

It is often presented as an alternative order to the figure of the State which, at best, is neither better nor worse and, in some cases, is even presented as being superior. Underlying all this is the idea of local residents accepting the authority of drug traffickers. Yet we must not forget that we are faced with a brutally imposed order, and we may well wonder why there is almost no account of the suffering caused by the traffickers' atrocities, or of the concerns linked to the control of certain spaces by armed factions, or of the dangerous horizons that bring so many young *favelados* into the ranks of violent gangs. By giving so little space to suffering and resistance, the tourism discourse also prolongs the law of silence. Perhaps what is even more serious is that this silence contributes to giving criminality the monopoly over resistance, which is a widespread trend of prejudice against the working classes. The figure of the criminal, who organizes an order under the nose of the State, is valorized due to their spectacular resistance to poverty. This resistance, however violent it may be, is easily interpreted in different ways within the frame of representations of those who occupy a higher social position. Violent and destructive resistance, in many respects, seems easier to understand than all the suffering that is kept quiet and the initiatives that are multiplied by the inhabitants.

Conclusion

This book has looked at the development of favela tourism from the perspective of the Western representations that condition it. The first chapter compared the representations at work in favela tours with other representations, constructed in particular through cinema, which show similar traits, one of the main ones being how the favela (in the singular) is often considered as a homogeneous category, which it is not in reality. The influence of cinematic representations of the favela does not summarize its place in the tourist industry. Rather, tourism and cinema expose their shared tendencies in how they use the image of the favela, and how they appear as the two modes of the favela's transformation into a symbol by Western society. Both modes take something located outside of Western society in order to include it within it. In many respects, the conditions of the favela's inclusion are fixed in advance: the favela is understood and valued according to preexisting categories, and according to the structures of Western apprehension of otherness. Western thought is taxonomic: it understands otherness only if it can order it, classify it and give it meaning. It is the conditions of this inclusion, experienced and modified in the course of touristic excursions (described in Chapters 1 and 2), that form the subject of this work.

The development of favela tourism thus provides a fertile ground for the study of Western constructions of otherness. This work argues that these constructions are mainly made in the mode of a rhetoric of authenticity, one I have named authenticism. Authenticity does indeed seem to be the central criterion for understanding and valorizating otherness. It is expressed in multiple ways in the case of favela tours. Indeed, Chapter 5 showed that the valorization of the favelas was mainly based on three elements (the favela as a place of preserved Brazilianness, as a space unexplored by tourists and as a territory of a culture of poverty), all three of which depend on the recognition of authenticity as a value to become positive elements.

Later on, in Chapter 6, I situated this authenticist posture in a longer history of the Western relationship with otherness. In doing so, I have introduced how the notion of exoticism, once dominant in the relationship of interest in the Other, and now distant in favor of authenticity, can be brought closer. Authenticity brings the apprehension of otherness into a regime of truth, not without the scientific influence of anthropology, and in this way participates in the negation of an asymmetrical relationship in the definition of the Other and in the selection of valued cultural traits. It conceals the way in which the construction of these positive representations leads to the taming, and even to the assimilation of otherness according to the terms that make it intelligible, in conformity with the categories formed to regulate relations to otherness and to construct an image of the world. The world constructed by and for tourism has symbolic uses in Western society. In Chapter 4, I mentioned the role played by strategies of distinction in these modes of consumption of otherness, in these "uses of the world", to use Nicolas Bouvier's expression (1985) in different terms. This symbolic charge, positive or negative, that mobility practices can take on has a considerable influence on the structuring of the offer in the favelas and in the operations of classification that tourists carry out on the basis of this offer. For the favela, these strategies show its place in what we could call an economy of otherness (Chapter 7). Tourism, in this way, creates a market of symbolic goods that tends to reduce otherness by transforming it into a symbol.

In the last two chapters of the book, I returned to the ethnography of the excursions in order to observe how, in the context of tourist consumption in the favela as a property of otherness, the interaction between tourists and the favela takes place. In order to compensate for the absence of encounters, that is, direct verbal interactions between tourists and inhabitants, I proposed in Chapter 8 an ethnography of gazes. Indeed, understanding the rules, explicit or tacit, which structure the conditions of exchanging gazes between tourists and inhabitants, enables us to see how the tourists' presence in the favelas is regulated. Moreover, the ethnography reveals the existence of a legitimate space of the gaze and of spaces (sometimes made of physical borders, but not entirely) where gazing is transgressive. Now, it appears that tourists frequently situate authenticity as located in this space whose perception, even if fleeting, allows them to take a glimpse of a world uninfluenced by their presence and not conditioned by the guide's official discourse. Continuing the analysis of the tourist experience of authenticity, Chapter 9 is devoted to the notion of reality. It shows the importance, following the ethnography of the gaze, of the apposition on the favela of a framework collectively formed for its tourist experience. Indeed, even though the reality is sought at all costs by the tourists, the visits to the favelas cannot escape the logics induced by the representations constructed for the tourist experience, and thus to the narrative framework. From this framework, and through the capital influence of the guide, a selection of the elements made visible during the visit takes place, others being simultaneously discarded. Moreover, the rhetoric of authenticity also influences the framework of the tours, notably in terms of interactions. Indeed, if a logic of avoidance dominates between tourists and locals, it is partly because of the importance of authenticity. On the one hand, tourists avoid the inhabitants in order to preserve authenticity by maintaining a reality uninfluenced by their presence. The structuring of practices by the rhetoric of authenticity transforms visits into an almost exclusively visual experience, and renders superfluous the idea of an encounter between tourists and inhabitants. On the side of the latter, the tourist frame is not very inclusive and is mostly left to the guide to mediate alone. In fact, the guides are those who have managed to capture the benefits of tourist mediation thanks to their mastery of Western codes (Chapter 3). In most cases, locals ignore tourists, avoiding talking to them and even, in some situations, looking at them. When they do interact with a group, it is by addressing the guide and, as I have shown, the guide is then free to translate the local's intervention into terms that avoid a disruption in the tourist frame.

Although favela tourism has existed since the early 1990s, it is still a new phenomenon, which experienced an initial boom in the 2010s before going through a significant decline as of 2016. In many aspects, it is too early to say what the birth of the tourist favela can bring to the favelas, which remain stigmatized spaces plagued by violence. Other studies have analyzed the economic consequences of tourism in the favelas (Duarte 2010; Rezende 2014) and questioned its impact (Frisch 2012; Duchesneau-Custeau 2020). This book, focusing on the tourists, has adopted a different angle. While its subject is not the impact of tourism on local development, the analyses of the symbolic frameworks of tourist practices nevertheless provide elements that show how it is also in the very substance of favela tours that we can look for certain limits of development through tourism, and in particular of the stated objective of destigmatizing the favela. Indeed, if the valorization by tourists of their place in life can be appreciated by the inhabitants (Freire-Medeiros 2010), it is certain that the tourist narratives that transform the favela into a symbolic good cannot totally lead to a change in its status. In summary, favela tourism is based on a symbolic inversion of the relationship with marginality and urban poverty, whose spaces pass thus by tourism, from a distressing symbol to the incarnation of an authentic otherness. Indeed, it is precisely the aspects that were obstacles to the presence of tourism (bad reputation, lack of infrastructure, danger) that have been transformed into selling points. In this respect, the historical stigmatization of the favela produces a guarantee of authenticity as the evolution of the view of the global modernization process and the evocation of the practices of leisure mobility (increase in the degree of independence, information and competence of tourists) have led to the emergence of a symbolic context in which the separation of the favela from these processes now appears, to travelers at least, as a preservation. In this way, it is the marginalization of the favelas, both in the territory of global tourism and on the more local scale of their political and social sidelining, that has served as a guarantee of authenticity in the context of their tourism over the last 20 years. In a concrete way, this nature of the tourist representations of the favela implies, for the guides, to insist on the constitutive aspects of marginalization, and to reintroduce them at the heart of their marketing and attraction discourses towards the tourists. It is in this way that the fundamental paradox of this frame in tourism is created. Marginal, therefore authentic, is the rhetoric of favela tourism. In these terms, it is not certain that the touristic representations of the favela constructed in a Western relationship with otherness, which this work has sought to analyze, and which are not exported to Brazilian society, can serve as a catalyst for a true destigmatization of favelas.

References

Abreu, M. (1994). Reconstruire une histoire oubliée : origine et expansion initiale des *favelas* de Rio de Janeiro. *Genèses*, 16, 45–68.

Affergan, F. (1987). *Exotisme et altérité. Essai sur les fondements d'une critique de l'anthropologie*. Presses universitaires de France, Paris.

Amselle, J.-L. (2010). *Rétrovolutions : essais sur les primitivismes contemporains*. Stock, Paris.

Araujo de Paula, L. (2011). L'évolution du phénomène des "milices" à Rio de Janreiro : une démocratie "en uniforme" ? *Dépasser les dichotomies : penser autrement les Amériques ?*, Paris.

Augé, M. (1994). *Pour une anthropologie des mondes contemporains*. Aubier, Paris.

Augé, M. (1997). *L'impossible voyage : le tourisme et ses images*. Rivages, Paris.

Bachelard, G. (1957). *La poétique de l'espace*. Presses universitaires de France, Paris.

Baudrillard, J. and Guillaume, M. (1992). *Figures de l'altérité*. Éditions Descartes, Paris.

Benjamin, W. (2011). *L'œuvre d'art à l'heure de sa reproductibilité technique*. Éditions Allia, Paris.

Bentes, I. (2003). The sertão and the favela in contemporary Brazilian film. In *The New Brazilian Cinema*, Nagib, L. (ed.). Tauris, London.

Berner, E. (1996). Legalizing squatters, excluding the poorest: Urban land transfer programs in the Philippines. Working Document, Bielefeld University.

Bertho-Lavenir, C. (1999). *La roue et le stylo : comment nous sommes devenus touristes ?* Odile Jacob, Paris.

Blunden, H. (2007). The impacts of the Sydney Olympic Games on housing rights. Background paper, COHRE, Geneva.

Boirot, A. (2020). Là où vont les garçons : une anthropologie du tourisme festif (Lloret de Mar, Costa Brava). Thesis, EHESS, Paris.

Borius, O. (2010). Rio de Janeiro olympique : murs tropicaux, favelas et gentrification. *Independencias – Depedencias – Interdependencias, VI Congresso CEISAL*, Toulouse.

Bourdieu, P. (1979). *La distinction : critique sociale du jugement*. Éditions de Minuit, Paris.

Bourdieu, P. (1984). *Distinction: A Social Critique of the Judgement of Taste*. Harvard University Press, Cambridge.

Bourdieu, P. (1997). *Les méditations pascaliennes*. Le Seuil, Paris.

Bourdieu, P. (2000). *Pascalian Meditations*. Stanford University Press.

Boutroy, E. (2006). Cultiver le danger dans l'alpinisme himalayen. *Ethnologie française*, 36, 591–601.

Bouvier, N. (1985). *L'usage du monde*. La Découverte, Paris.

Boyer, M. (1999). *Histoire du tourisme de masse*. Presses universitaires de France, Paris.

Boyer, M. (2000). *Histoire de l'invention du tourisme, XVIe–XIXe siècles : origine et développement du tourisme dans le sud-est de la France*. Éditions de l'Aube, La Tour-d'Aigue.

Boyer, M. (2002). Comment étudier le tourisme. *Ethnologie française*, 32, 393–404.

Broudehoux, A.M. (2017). *Mega-events and Urban Image Construction. From Beijing to Rio de Janeiro*. Routledge, London.

Brown, D. (1999). De faux authentiques. *Terrain*, 33, 41–56.

Bruner, E. (1989). Of cannibals, tourists, and ethnographers, review of "Cannibal Tours" by Denis O'Rourke. *Cultural Anthropology*, 4, 438–445.

Bruner, E. (2005). *Culture on Tour: Ethnographies of Travel*. University of Chicago Press.

Burgold, J. and Rolfes, M. (2013). Of voyeuristic safari tours and responsible tourism with educational value: Observing moral communication in slum and township tourism in Cape Town and Mumbai. *Journal of the Geographical Society of Berlin*, 144(2), 161–174.

Butcher, J. (2003). *The Moralisation of Tourism: Sun, Sand... and Saving the World?* Routledge Contemporary Geographies of Leisure, London.

de Certeau, M. (1985). Où ça commence. Fondation d'histoire. Les mystiques du XVIe et du XVIIe siècle. *Mi-dit*, 10(11), 6–13.

Chabloz, N. (2006). Vers une éthique du tourisme ? *Autrepart*, 40, 45–62.

Chege, P. and Mwisukha, A. (2013). Benefits of slum tourism in Kibera slum in Nairobi, Kenya. *International Journal of Arts and Commerce*, 2(4), 94–102.

Cohen, E. (1972). Toward a sociology of international tourism. *Social Research*, 39(1), 164–182.

Cohen, E. (1985). The tourist guide: The origin, structure and dynamics of a role. *Annals of Tourism Research*, 12, 5–29.

Cohen, E. (2012). Authentification: Hot and cool. *Annals of Tourism Research*, 39, 1295–1314.

Cornelissen, S., Swart, K., Bob, U. (2011). Sport mega-events and their legacy: The 2010 FIFA World Cup. *Development Southern Africa*, 28(3), 305–306.

Coulangeon, P. (2004). Classes sociales, pratiques culturelles et styles de vie. Le modèle de la distinction est-il (vraiment) obsolète ? *Sociologie et sociétés*, 36(1), 59–85.

Cousin, S. (2011). Destination authentique : le tourisme, ou la quête (é)perdue de l'authenticité. *Les cahiers du Musée des Confluences*, 8, 59–66.

Cousin, S. (2016). Imaginaries and imagination. French notes. In *Tourism Imaginaries at the Disciplinary Crosswords: Place, Practice, Media*, Graburn, N. and Gravari-Barbas, M. (eds). Routledge, London.

Cousin, S. (2018). Économies de l'altérité. Dossier pour l'habilitation à diriger des recherches, ENS, PSL, Paris.

Cousin, S. (2020). Réinventons les hétérotopies du tourisme. *Espaces*, 352, 71–75.

Cousin, S. and Apchain, T. (2016). Tourisme et anthropologie : un tango de l'altérité. *Mondes du Tourisme* [Online]. Available at: http://journals.openedition.org/tourisme/1320 [Accessed 1 September 2021].

Cousin, S. and Réau, B. (2009). *Sociologie du tourisme*. La Découverte, Paris.

Cravatte, C. (2009). L'anthropologie du tourisme et l'authenticité. Catégorie analytique ou catégorie indigène ? *Cahiers d'études africaines*, 193(194), 603–319.

Cunha, E. (1985). *Os sertões: Campanha de Canudos (1902)*. Brasiliense, São Paulo.

Davis, L.K. (2007). *Housing, Evictions, and the Seoul 1988 Summer Olympic Games*. COHRE, Genève.

Deluchey, J.F. (2003). De la "guerre contre le crime" au Brésil : culture autoritaire et politiques publiques de la sécurité. *Autre-part*, 2(46), 173–186.

Di Maggio, P. (1987). Classification in art. *American Sociological Review*, 52(4), 440–455.

Diderot, D. (2006). *Supplément au "Voyage de Bougainville"*. Flammarion, Paris.

Diekmann, A. and Hannam, K. (2012). Touristic mobilities in India's slum spaces. *Annals of Tourism Research*, 39(3), 1315–1336.

Dilthey, W. (1976). *Dilthey: Selected Writings*. Cambridge University Press.

Doquet, A. (2009). Guides, guidons et guitares. Authenticité et guides touristiques au Mali. *Cahiers d'études africaines*, 193(194), 73–74.

Duarte, R. (2010). Exploring the social impact of favela tourism. An insight into the residents' view. Master's thesis, Wageningen.

Duchesneau-Custeau, A. (2020). "Theirs seemed like a lot more fun": Favela tourism, commodization of poverty and stereotypes of the self and of the other. Research document, School of International Development and Global Studies, University of Ottawa.

Durkheim, E. (1930). *De la division du travail social*. Presses universitaires de France, Paris.

Durr, E. (2012). Urban poverty, spatial representation and mobility. Touring a slum in Mexico. *International Journal of Urban and Regional Research*, 36(4), 706–724.

Durr, E. and Jaffe, R. (2012). Theorizing slum tourism: Performing, negotiating and transforming inequality. *European Review of Latin American and Caribbean Studies*, 93, 113–123.

Duval, J. (2020). Espace social. In *Dictionnaire International Bourdieu*, Sapiro, G. (ed.). CNRS Éditions, Paris.

Duvoux, N. (2010). Repenser la culture de pauvreté. *La vie des idées* [Online]. Available at: https://laviedesidees.fr/Repenser-la-culture-de-la-pauvrete.html [Accessed 24 February 2022].

Dyson, P. (2012). Slum tourism: Representing and interpreting "reality" in Dharavi, Mumbai. *Tourism Geographies*, 14(2), 254–274.

Élias, N. (1991). Conscience de soi et image de soi. In *La société des individus*, Elias, N. (ed.). Fayard, Paris.

Fléchet, A. (2008). L'exotisme comme objet d'histoire. *Hypothèses*, 11, 15–26.

Foucault, M. (1976). *Surveiller et punir : naissance de la prison*. Gallimard, Paris.

Foucault, M. (1991). *The Birth of the Clinic: An Archaeology of Medical Perception*. Routledge, London.

Foucault, M. (ed.) (2001). Des espaces autres. *Dits et écrits (1954–1988)*. Gallimard, Paris.

Foucault, M. (2003). *Naissance de la clinique*. Presses universitaires de France, Paris.

Freire-Medeiros, B. (2007). A favela que se vê e que se vende. Reflexões e polêmicas em torno de um destino turístico. *Revista Brasileira de Ciências Sociais*, 22(65), 62–72.

Freire-Medeiros, B. (2009). *Gringo na laje: produção, circulação e consumo da favela turística*. Editora FGV de Bolso, Rio de Janeiro.

Freire-Medeiros, B. (2010). Entre tapas e beijos: A favela turistica na perspectiva de seus moradores. *Revista Sociedade et Estado*, 25(1), 33–51.

Frenzel, F. (2016). *Slumming It. The Tourist Valorization of Urban Poverty*. Zed Books, London.

Frenzel, F. and Koens, K. (2012). Slum tourism: Developments in a young field of interdisciplinary tourism research. *Tourism Geographies*, 14(2), 195–212.

Frenzel, F., Koens, K., Steinbrinck, M. (2012). *Slum Tourism, Poverty, Power and Ethics*. Routledge, London.

Frisch, T. (2012). Glimpses of another world: The favela as tourist attraction. *Tourism Geographies*, 14(2), 195–212.

Garfinkel, H. (1967). *Recherches en ethnométhodologie*. Presses universitaires de France, Paris.

Goffman, E. (1974). *Frame Analysis: An Essay on the Organization of Experience*. Harvard University Press, Cambridge.

Goffman, E. (1979). *La mise en scène de la vie quotidienne*. Éditions de Minuit, Paris.

Gontard, M. (1990). *Victor Segalen : une esthétique de la différence*. L'Harmattan, Paris.

Goody, J. (1979). *La raison graphique*. Éditions de Minuit, Paris.

Graburn, N. (1977). Tourism: The sacred journey. In *Hosts and Guests*, Smith, V. (ed.). University of Pennsylvania Press.

Gravari-Barbas, M. and Jacquot, S. (2007). L'événement, outil de légitimation de projets urbains : l'instrumentalisation des espaces et des temporalités évenementiels à Lille et à Gênes. *Géocarrefour*, 82(3) [Online]. Available at: http://journals.openedition.org/geocarrefour/2217 [Accessed 9 February 2022].

Handler, R. and Linnekin, J. (1984). Tradition, genuine or spurious. *The Journal of American Folklore*, 97(385), 237–290.

Haroche, C. (2004). Manières de regarder dans les sociétés démocratiques contemporaines. Le sens du regard. *Communications*, 75, 147–169.

Haroche, C. and Vigarello, G. (2004). Le sens du regard. *Communications*, 75.

Heinich, N. (1997). *L'art contemporain exposé aux rejets : étude de cas*. Éditions Jacqueline Chambon, Nîmes.

Héritier, F. (2004). Regard et anthropologie. *Communications*, 75, 91–110.

Hobsbawn, E. and Ranger, T. (1983). *The Invention of Tradition*. Cambridge University Press.

Hoggart, R. (1957). *The Uses of Literacy*. Chatto and Windus, London.

Hoggart, R. (1970). *La culture du pauvre : étude sur le style de vie des classes populaires en Angleterre*. Les Éditions de Minuit, Paris.

Husserl, E. (2018). *Idées directrices pour une phénoménologie pure et une philosophie phénoménologique*. Gallimard, Paris.

Jaguaribe, B. (2004). Favelas and the aesthetics of realism: Representations in film and literature. *Journal of Latin American Cultural Studies*, 13(3), 327–342.

Karady, V. (1998). La république des lettres des temps modernes. L'internationalisation des marchés universitaires occidentaux avant la Grande Guerre. *Actes de la recherche en sciences sociales*, 121–122, 92–103.

Kirshenblatt-Gimblett, B. (1998). *Destination Culture: Tourism, Museums and Heritage*. University of California Press, Berkeley.

Koven, J. (2006). *Slumming: Sexual and Social Politics in Victorian London*. Princeton University Press.

Lahire, B. (2004). *La culture des individus : dissonances culturelles et distinction de soi*. La Découverte, Paris.

Leclerc, G. (ed.) (2006). *Le regard et le pouvoir*. Presses universitaires de France, Paris.

Leclerc, G. (2015). *Le désir de voyage et la quête de l'Autre*. L'Harmattan, Paris.

Leite, N. and Graburn, N. (2010). L'anthropologie pour étudier le tourisme. *Mondes du tourisme*, 1, 17–28.

Lenclud, G. (1996). Le grand partage ou la tentation ethnologique. In *Vers une ethnologie du présent*, Althabe, G., Fabre, D., Lenclud, G. (eds). Éditions de la Maison des sciences de l'homme, Paris.

Leopold, A. (1949). *A Sand County Almanac*. Oxford University Press.

Leopold, A. (2000). *Almanach d'un comté des sables*. Flammarion, Paris.

Leu, L. (2004). Fantasia e fetiche: Consomindo o Brasil na Ingleterra. *Eco-Pos*, 7(2), 13–72.

Lévi-Strauss, C. (1963). *Structural Anthropology*. Basic Books.

Lewis, O. (1959). *Five Families: Mexican Case Studies in the Culture of Poverty*. Basic Books, New York.

Lewis, O. (1969). *La Vida. Une famille portoricaine dans une culture de pauvreté : San Juan et New York*. Gallimard, Paris.

Loti, P. (1899). *Madame Chrysanthème*. Calmann-Lévy, Paris.

Ma, B. (2010). A trip into the controversy: A study of slum tourism travel motivations. 2009–2010 Penn Humanities Forum on Connections.

MacCannell, D. (1973). Staged authenticity: On arrangements of social space in tourist settings. *The American Journal of Sociology*, 79(3), 589–603.

MacCannell, D. (1976). *The Tourist: A New Theory of the Leisure Class*. Schocken, New York.

Malinowski, B. (1963). *Les argonautes du Pacifique occidental*. Gallimard, Paris.

Malinowski, B. (2005). *Argonauts of the Western Pacific*. Taylor & Francis e-Library.

Maoz, D. (2006). The mutual gaze. *Annals of Tourism Research*, 33(1), 221–239.

Mason, P. (1998). *Infelicities. Representations of the Exotic*. The Johns Hopkins University Press, Baltimore.

Mathè, R. (1972). *L'exotisme : d'Homère à Le Clézio*. Bordas, Paris.

Mauss, M. (1936). Les techniques du corps. In *Sociologie et anthropologie*, Mauss, M. (ed.). Presses universitaires de France, Paris.

Mauss, M. (1973). The techniques of the body. *Economy and Society*, 2(1), 70–88.

Mehmetoglu, M. (2004). A typology of tourists from a different angle. *International Journal of Hospitality & Tourism Administration*, 5(3), 69–90.

Mekawy, M. (2012). Responsible slum tourism: Egyptian experience. *Annals of Tourism Research*, 39(4), 2092–2113.

Menezes, P.V. (2014). Os rumores da pacifição: A chegada da UPP e as mudanças nos problemos públicos no Santa Marta e na Cidade de Deus. *Dilemas, Revista de Estudos de Conflito e Controle Social*, 7(4), 665–684.

Meschkank, J. (2011). Investigations into slum tourism in Mumbai: Poverty tourism and the tensions between different constructions of reality. *GeoJournal*, 76, 47–62.

Morand, P. (1928). *Rien que la terre*. Grasset, Paris.

Moura, J.-M. (1992). *Lire l'exotisme*. Dunod, Paris.

Moura, J.-M. (1998). *La littérature des lointains : histoire de l'exotisme européen au XXe siècle*. Champion, Paris.

Nora, P. (1984). *Les lieux de mémoire*. Gallimard, Paris.

Novy, J. (2011). Marketing marginalized neighboroods. Tourism and leisure in the 21st century inner city. PhD Thesis, Columbia University, New York.

Olivier de Sardan, J.-P. (2008). *Anthropology and Development: Understanding Contemporary Social Change*. Bloomsbury Publishing, London.

Pagden, A. (1993). *European Encounters with the New World from Renaissance to Romanticism*. Yale University Press, New Haven.

Pasquier, D. (2005). *Cultures lycéennes : la tyrannie de la majorité*. Autrement, Paris.

Picard, D. (2011). *Tourism, Magic and Modernity: Cultivating the Human Garden*. Berghahn, New York.

Préteceille, E. and Valladares, L. (2000). Favela, favelas: Uncicidade ou diversidade da favela carioca. In *O futuro das metropoles: Desigualades e governabilidade*, de Queiroz, L.C. (ed.). Revan, Rio de Janeiro.

Racine, J. (ed.) (2017). Seconde préface de Bajazet. In *Théâtre Complet*. Éditions Classiques Garnier, Paris.

Réau, B. and Poupeau, F. (2007). L'enchantement du monde touristique. *Actes de la recherche en sciences sociales*, 170, 4–14.

Rezende, R. (2014). Turismo comunitario e favela-tour como expresses das novas dinámicas de consumo turístico. *Revista Turismo em Analise*, 25(2), 354–372.

Ringelblum, E. (1995). *Chroniques du ghetto de Varsovie*. Payot, Paris.

Roche, D. (2003). *Humeurs vagabondes. De la circulation des hommes et de l'utilité des voyages*. Fayard, Paris.

Rolfes, M. (2010). Poverty tourism: Theoretical reflections and empirical findings regarding an extraordinary form of tourism. *GeoJournal*, 75(5), 421–442.

Rolfes, M., Steinbrinck, M., Uhl, C. (2009). Townships as attraction: An empirical study of township tourism in Cape Town. Research report, University of Potsdam.

Rushdie, S. (1991). *Imaginary Homelands. Essays and Criticism 1981–1991*. Granta, London.

Sapir, E. (1967). *Anthropologie*. Éditions de Minuit, Paris.

Sartre, J.-P. (1984). *Being and Nothingness*. Philosophical Library.

Sartre, J.-P. (1989). *No Exit, and Three Other Plays*. Vintage International, New York.

Schopenhauer, A. (2004). *Le monde comme volonté et représentation*. Presses universitaires de France, Paris.

Schutz, A. (1962). *Collected Papers I: The Problem of Social Reality*. Martinus Nijhoff, The Hague.

Segalen, V. (1976). *Les origines de la statuaire de Chine*. Éditions de la Différence, Paris.

Segalen, V. (1978). *Essai sur l'exotisme : une esthétique du divers*. Éditions Fata Morgana, Montpellier.

Segalen, V. (2021). *Essai sur le Mystérieux et autres textes*. Rivages poche, Paris.

Selwyn, T. (1996). Introduction. In *The Tourism Image: Myths and Myth Making in Tourism*, Selwyn, T. (ed.). John Wiley & Sons Ltd., London.

Semmel, G. (1994). Bridge and door. *Theory, Culture and Society*, 11, 5–10.

Sharpley, R. (1994). *Tourism, Tourists and Society*. ELM, Huntingdon.

Shinn, H.B. (2009). Life in the shadow of mega-events: Beijing Summer Olympiad and its impact on housing. *Journal of Asian Public Policy*, 2(2), 122–141.

Simmel, G. (1988). Pont et porte. In *La tragédie de la culture*, Simmel, G. (ed.). Rivages, Paris.

Simmel, G. (1999). Excursus sur la sociologie des sens. In *Sociologie, études sur les formes de socialisation*, Simmel, G. (ed.). Gallimard, Paris.

Small, M.L., Harding, D.J., Lamont, M. (2010). Reconsidering culture and poverty. *The Annals of the American Academy of Political and Social Science*, 629(1), 6–27.

Soares Goncalves, R. (2010). *Les favelas de Rio de Janeiro, histoire et droit : XIXe–XXe siècles*. L'Harmattan, Paris.

Sontag, S. (2008). *On Photography*. Penguin, Harmondsworth.

Staszak, J.-F. (2008). Qu'est-ce que l'exotisme ? *Le Globe. Revue genevoise de géographie*, 148, 7–30.

Stone, P. and Sharpley, R. (2008). Consuming dark tourism. A thanatological perspective. *Annals of Tourism Research*, 35(2), 574–595.

Todorov, T. (1989). *Nous et les autres : la réflexion française sur la diversité humaine*. Le Seuil, Paris.

Urbain, J.-D. (1991). *L'idiot du voyage : histoires de touristes*. Plon, Paris.

Urbain, J.-D., Doquet, A., Evrard, O. (2008). An interview with Jean-Didier Urbain: Tourism beyond the grave: A semiology of culture. *Tourist Studies*, 8(2), 175–191.

Urry, J. (1990). *The Tourist Gaze: Leisure and Travel in Contemporary Societies*. Sage, London.

Urry, J. and Larsen, J. (2011). *The Tourist Gaze 3.0*. Sage, London.

Vainer, C. (2011). Cidade de Exceção: Reflexões a Partir do Rio de Janeiro. *Anais do XIV Encontró Nacional da Associação de Plajamento Urbano (Anpur)*, 14.

Valladares, L. (2006). *La favela d'un siècle à l'autre*. Éditions de la Maison des sciences de l'homme, Paris.

Veblen, T. (1899). *The Theory of the Leisure Class: An Economic Study of Institutions*. The Macmillan Company, New York.

Wachsberger, M. (2008). La favela d'un siècle à l'autre. *Population*, 63(1), 210–211.

Wagner, A.-C. (2007). La place du voyage dans la formation des élites. *Actes de la recherche en sciences sociales*, 170, 61–67.

Wang, N. (1999). Rethinking authenticity in tourism experience. *Annals of Tourism Research*, 26(2), 249–370.

Wang, H., Xiaokaiti, M., Zhou, Y., Yang, Y., Liu, Y., Zhao, R. (2012). Mega-events and city branding: A case study of Shanghai World Expo 2010. *Journal of US-China Public Administration*, 9(11), 1283–1293.

Warnier, J.-P. (ed.) (1994). *Le paradoxe de la marchandise authentique*. L'Harmattan, Paris.

Williamson, T. (2011). Not everyone has a price: How the small favela of Vila Autodromo's fight opened a path to Olympic resistance. In *Rio 2016: Olympic Myths, Hard Realities*, Zimbalist, A. (ed.). Brookings Institution Press, Washington.

Wilson, W.J. (1987). *The Truly Disadvantaged. The Inner City, the Underclass, and Public Policy*. University of Chicago Press.

Zaluar, A. and Conceição, I. (2007). Favelas sob o controle das milicias no Rio de Janeiro. Que paz? *São Paulo em perspectiva*, 21(2), 89–101.

Zimbalist, A. (2017). The economic legacy of Rio 2016. In *Rio 2016: Olympic Myths, Hard Realities*, Zimbalist, A. (ed.). Brookings Institution Press, Washington.

Index

Other titles from

ISTE

in

Science, Society and New Technologies

2023

CARDON Alain, EL HAMI Abdelkhalak
Fundamental Generation Systems: Computer Science and Artificial Consciousness, the Informational Field of Generation of the Universe, the Sixth Sense of Living Beings
(Digital Sciences Set – Volume 4)

ELAMÉ Esoh
The Sustainable City in Africa Facing the Challenge of Liquid Sanitation
(Territory Development Set – Volume 2)

JURCZENKO Emmanuel
Climate Investing: New Strategies and Implementation Challenges

PARET Dominique, CRÉGO Pierre, SOLÈRE Pauline
Smart Patches: Biosensors, Graphene and Intra-Body Communications

VENTRE Daniel, LOISEAU Hugo
Cybercrime during the SARS-CoV-2 Pandemic (2019-2022): Evolutions, Adaptations, Consequences
(Cybersecurity Set – Volume 3)

2022

AIT HADDOU Hassan, TOUBANOS Dimitri, VILLIEN Philippe
Ecological Transition in Education and Research

CARDON Alain
Information Organization of The Universe and Living Things: Generation of Space, Quantum and Molecular Elements, Coactive Generation of Living Organisms and Multiagent Model
(Digital Science Set – Volume 3)

CAULI Marie, FAVIER Laurence, JEANNAS Jean-Yves
Digital Dictionary

DAVERNE-BAILLY Carole, WITTORSKI Richard
Research Methodology in Education and Training: Postures, Practices and Forms
(Education Set – Volume 12)

ELAMÉ Esoh
Sustainable Intercultural Urbanism at the Service of the African City of Tomorrow
(Territory Development Set – Volume 1)

FLEURET Sébastien
A Back and Forth Between Tourism and Health: From Medical Tourism to Global Health
(Tourism and Mobility Systems Set – Volume 5)

KAMPELIS Nikos, KOLOKOTSA Denia
Smart Zero-energy Buildings and Communities for Smart Grids
(Engineering, Energy and Architecture Set – Volume 9)

2021

BARDIOT Clarisse
Performing Arts and Digital Humanities: From Traces to Data
(Traces Set – Volume 5)

BENSRHAIR Abdelaziz, BAPIN Thierry
From AI to Autonomous and Connected Vehicles: Advanced Driver-Assistance Systems (ADAS)
(Digital Science Set – Volume 2)

BÉRANGER Jérôme
Societal Responsibility of Artificial Intelligence: Towards an Ethical and Eco-responsible AI
Technological Prospects and Social Applications Set – Volume 4)

CORDELIER Benoit, GALIBERT Oliver
Digital Health Communications
Technological Prospective and Social Applications Set – Volume 5)

DOUAY Nicolas, MINJA Michael
Urban Planning for Transitions

GALINON-MÉLÉNEC Béatrice
The Trace Odyssey 1: A Journey Beyond Appearances
(Traces Set – Volume 4)

HENRY Antoine
Platform and Collective Intelligence: Digital Ecosystem of Organizations

KOLOKOTSA Denia, KAMPELIS Nikos
Smart Buildings, Smart Communities and Demand Response
(Engineering, Energy and Architecture Set – Volume 8)

LE LAY Stéphane, SAVIGNAC Emmanuelle, LÉNEL Pierre, FRANCES Jean
The Gamification of Society
(Research, Innovative Theories and Methods in SSH Set – Volume 2)

RADI Bouchaïb, EL HAMI Abdelkhalak
Optimizations and Programming: Linear, Non-linear, Dynamic, Stochastic and Applications with Matlab
(Digital Science Set – Volume 1)

2020

ALAKTIF Jamila, CALLENS Stéphane
Migration and Climate Change: From the Emergence of Human Cultures to Contemporary Management in Organizations

BARNOUIN Jacques
The World's Construction Mechanism: Trajectories, Imbalances and the Future of Societies
(Interdisciplinarity between Biological Sciences and Social Sciences Set – Volume 4)

ÇAĞLAR Nur, CURULLI Irene G., SIPAHIOĞLU Işıl Ruhi, MAVROMATIDIS Lazaros
Thresholds in Architectural Education (Engineering, Energy and Architecture Set – Volume 7)

DUBOIS Michel J.F.
Humans in the Making: In the Beginning was Technique
(Social Interdisciplinarity Set – Volume 4)

ETCHEVERRIA Olivier
The Restaurant, A Geographical Approach: From Invention to Gourmet Tourist Destinations
(Tourism and Mobility Systems Set – Volume 3)

GREFE Gwenaëlle, PEYRAT-GUILLARD Dominique
Shapes of Tourism Employment: HRM in the Worlds of Hotels and Air Transport (Tourism and Mobility Systems Set – Volume 4)

JEANNERET Yves
The Trace Factory
(Traces Set – Volume 3)

KATSAFADOS Petros, MAVROMATIDIS Elias, SPYROU Christos
Numerical Weather Prediction and Data Assimilation (Engineering, Energy and Architecture Set – Volume 6)

MARTI Caroline
Cultural Mediations of Brands: Unadvertization and Quest for Authority
(Communication Approaches to Commercial Mediation Set – Volume 1)

MAVROMATIDIS Lazaros E.
Climatic Heterotopias as Spaces of Inclusion: Sew Up the Urban Fabric
(Research in Architectural Education Set – Volume 1)

MOURATIDOU Eleni
Re-presentation Policies of the Fashion Industry: Discourse, Apparatus and Power
(Communication Approaches to Commercial Mediation Set – Volume 2)

SCHMITT Daniel, THÉBAULT Marine, BURCZYKOWSKI Ludovic
Image Beyond the Screen: Projection Mapping

VIOLIER Philippe, with the collaboration of TAUNAY Benjamin
The Tourist Places of the World
(Tourism and Mobility Systems Set – Volume 2)

2019

BRIANÇON Muriel
The Meaning of Otherness in Education: Stakes, Forms, Process, Thoughts and Transfers
(Education Set – Volume 3)

DESCHAMPS Jacqueline
Mediation: A Concept for Information and Communication Sciences
(Concepts to Conceive 21st Century Society Set – Volume 1)

DOUSSET Laurent, PARK Sejin, GUILLE-ESCURET Georges
Kinship, Ecology and History: Renewal of Conjunctures
(Interdisciplinarity between Biological Sciences and Social Sciences Set – Volume 3)

DUPONT Olivier
Power
(Concepts to Conceive 21st Century Society Set – Volume 2)

FERRARATO Coline
Prospective Philosophy of Software: A Simondonian Study

GUAAYBESS Tourya
The Media in Arab Countries: From Development Theories to Cooperation Policies

HAGÈGE Hélène
Education for Responsibility
(Education Set – Volume 4)

LARDELLIER Pascal
The Ritual Institution of Society
(Traces Set – Volume 2)

LARROCHE Valérie
The Dispositif
(Concepts to Conceive 21st Century Society Set – Volume 3)

LATERRASSE Jean
Transport and Town Planning: The City in Search of Sustainable Development

LELEU-MERVIEL Sylvie, SCHMITT Daniel, USEILLE Philippe
From UXD to LivXD: Living eXperience Design

LENOIR Virgil Cristian
Ethically Structured Processes
(Innovation and Responsibility Set – Volume 4)

LOPEZ Fanny, PELLEGRINO Margot, COUTARD Olivier
Local Energy Autonomy: Spaces, Scales, Politics
(Urban Engineering Set – Volume 1)

METZGER Jean-Paul
Discourse: A Concept for Information and Communication Sciences
(Concepts to Conceive 21st Century Society Set – Volume 4)

MICHA Irini, VAIOU Dina
Alternative Takes to the City
(Engineering, Energy and Architecture Set – Volume 5)

PÉLISSIER Chrysta
Learner Support in Online Learning Environments

PIETTE Albert
Theoretical Anthropology or How to Observe a Human Being
(Research, Innovative Theories and Methods in SSH Set – Volume 1)

PIRIOU Jérôme
The Tourist Region: A Co-Construction of Tourism Stakeholders
(Tourism and Mobility Systems Set – Volume 1)

PUMAIN Denise
Geographical Modeling: Cities and Territories
(Modeling Methodologies in Social Sciences Set – Volume 2)

WALDECK Roger
Methods and Interdisciplinarity
(Modeling Methodologies in Social Sciences Set – Volume 1)

2018

BARTHES Angela, CHAMPOLLION Pierre, ALPE Yves
Evolutions of the Complex Relationship Between Education and Territories
(Education Set – Volume 1)

BÉRANGER Jérôme
The Algorithmic Code of Ethics: Ethics at the Bedside of the Digital Revolution
(Technological Prospects and Social Applications Set – Volume 2)

DUGUÉ Bernard
Time, Emergences and Communications
(Engineering, Energy and Architecture Set – Volume 4)

GEORGANTOPOULOU Christina G., GEORGANTOPOULOS George A.
Fluid Mechanics in Channel, Pipe and Aerodynamic Design Geometries 1
(Engineering, Energy and Architecture Set – Volume 2)

GEORGANTOPOULOU Christina G., GEORGANTOPOULOS George A.
Fluid Mechanics in Channel, Pipe and Aerodynamic Design Geometries 2
(Engineering, Energy and Architecture Set – Volume 3)

GUILLE-ESCURET Georges
Social Structures and Natural Systems: Is a Scientific Assemblage Workable?
(Social Interdisciplinarity Set – Volume 2)

LARINI Michel, BARTHES Angela
Quantitative and Statistical Data in Education: From Data Collection to Data Processing
(Education Set – Volume 2)

LELEU-MERVIEL Sylvie
Informational Tracking
(Traces Set – Volume 1)

SALGUES Bruno
Society 5.0: Industry of the Future, Technologies, Methods and Tools
(Technological Prospects and Social Applications Set – Volume 1)

TRESTINI Marc
Modeling of Next Generation Digital Learning Environments: Complex Systems Theory

2017

ANICHINI Giulia, CARRARO Flavia, GESLIN Philippe,
GUILLE-ESCURET Georges
Technicity vs Scientificity – Complementarities and Rivalries
(Interdisciplinarity between Biological Sciences and Social Sciences Set – Volume 2)

DUGUÉ Bernard
Information and the World Stage – From Philosophy to Science, the World of Forms and Communications
(Engineering, Energy and Architecture Set – Volume 1)

GESLIN Philippe
Inside Anthropotechnology – User and Culture Centered Experience
(Social Interdisciplinarity Set – Volume 1)

GORIA Stéphane
Methods and Tools for Creative Competitive Intelligence

KEMBELLEC Gérald, BROUDOUS EVELYNE
Reading and Writing Knowledge in Scientific Communities: Digital Humanities and Knowledge Construction

MAESSCHALCK Marc
Reflexive Governance for Research and Innovative Knowledge (Responsible Research and Innovation Set - Volume 6)

PARK Sejin, GUILLE-ESCURET Georges
Sociobiology vs Socioecology: Consequences of an Unraveling Debate (Interdisciplinarity between Biological Sciences and Social Sciences Set – Volume 1)

PELLÉ Sophie
Business, Innovation and Responsibility (Responsible Research and Innovation Set – Volume 7)

2016

ANDRÉ Michel, SAMARAS Zissis
Energy and Environment (Research for Innovative Transports Set – Volume 1)

BLANQUART Corinne, CLAUSEN Uwe, JACOB Bernard
Towards Innovative Freight and Logistics (Research for Innovative Transports Set – Volume 2)

BRONNER Gérald
Belief and Misbelief Asymmetry on the Internet

COHEN Simon, YANNIS George
Traffic Management (Research for Innovative Transports Set – Volume 3)

EL FALLAH SEGHROUCHNI Amal, ISHIKAWA Fuyuki, HÉRAULT Laurent, TOKUDA Hideyuki
Enablers for Smart Cities

GIANNI Robert
Responsibility and Freedom (Responsible Research and Innovation Set – Volume 2)

GRUNWALD Armin
The Hermeneutic Side of Responsible Research and Innovation
(Responsible Research and Innovation Set – Volume 5)

LAGRAÑA Fernando
E-mail and Behavioral Changes: Uses and Misuses of Electronic Communications

LENOIR Virgil Cristian
Ethical Efficiency: Responsibility and Contingency
(Responsible Research and Innovation Set – Volume 1)

MAESSCHALCK Marc
Reflexive Governance for Research and Innovative Knowledge
(Responsible Research and Innovation Set – Volume 6)

PELLÉ Sophie, REBER Bernard
From Ethical Review to Responsible Research and Innovation
(Responsible Research and Innovation Set – Volume 3)

REBER Bernard
Precautionary Principle, Pluralism and Deliberation: Sciences and Ethics
(Responsible Research and Innovation Set – Volume 4)

TORRENTI Jean-Michel, LA TORRE Francesca
Materials and Infrastructures 1
(Research for Innovative Transports Set – Volume 5A

TORRENTI Jean-Michel, LA TORRE Francesca
Materials and Infrastructures 2
(Research for Innovative Transports Set – Volume 5B)

VENTRE Daniel
Information Warfare – 2nd edition

YANNIS George, COHEN Simon
Traffic Safety
(Research for Innovative Transports Set – Volume 4)

www.ingramcontent.com/pod-product-compliance
Ingram Content Group UK Ltd.
Pitfield, Milton Keynes, MK11 3LW, UK
UKHW020811290926
15470UKWH00003B/1079

9 781786 307675